Greenlanders, Whales, and Whaling

ARCTIC VISIONS

Gail Osherenko and Oran Young

GENERAL EDITORS

The Arctic has long appeared to outsiders as a vast, forbidding wasteland or, alternatively, as a storehouse of riches ready for the taking by those able to conquer the harsh physical environment. More recently, a competing vision paints the Arctic as the last pristine wilderness on earth, a place to be preserved for future generations.

Arctic Visions confronts these conflicting and simplistic portraits, conceived in ignorance of the complexities of the circumpolar world and without appreciation of the viewpoints of those indigenous to the region. Drawing upon an international community of writers who are sensitive to human dimensions, Arctic Visions will explore political, strategic, economic, environmental, and cultural issues.

The Arctic has always been a place of human and natural drama, an arena for imperial ambitions, economic exploitation, ecological disasters, and personal glory. As the region gains importance in international affairs, this series will help a growing audience of readers to develop new and more informed visions of the Arctic.

Arctic Politics: Conflict and Cooperation in the Circumpolar North, Oran Young, 1992

Arctic Wars, Animal Rights, Endangered Peoples, Finn Lynge, 1992

Arctic Adaptations: Native Whalers and Reindeer Herders of Northern Eurasia, Igor Krupnik, 1993

Relocating Eden: The Image and Politics of Inuit Exile in the Canadian Arctic, Alan Rudolph Marcus, 1995

Sold American: The Story of Alaska Natives and Their Land, 1867–1959: The Army to Statehood, Donald Craig Mitchell, 1997

Greenlanders, Whales, and Whaling: Sustainability and Self-Determination in the Arctic, Richard A. Caulfield, 1997

Greenlanders, Whales, and Whaling

SUSTAINABILITY AND SELF-DETERMINATION IN THE ARCTIC

Richard A. Caulfield

DARTMOUTH COLLEGE

Published by University Press of New England / Hanover and London

DARTMOUTH COLLEGE

Published by University Press of New England,
Hanover, NH 03755
© 1997 by the Trustees of Dartmouth College
All rights reserved
Printed in the United States of America

5 4 3 2 1

CIP data appear at the end of the book

for my wife, Annie,
and for Caitlin, Michael, and Julia
who shared the journey with me

Contents

Illustrations and Tables

Tables

Acknowledgments

This book could not have been written without the approval and cooperation of the people of Qeqertarsuaq Municipality, West Greenland. I thank *borgmesters* Augusta Salling and Jens Johan Broberg and the leadership of KNAPP, the local hunters and fishers association, for their kind assistance. I offer a special note of appreciation to elders Hans Egede Berthelsen, the late Katrine Berthelsen, David Broberg, Esais Broberg, Lars Broberg, the late Hansipaluk Broberg, Albrecht Olrik, and Adam Wille.

In particular, I want to express my appreciation for the extraordinary contributions of Lars Pele and Magdalene Berthelsen, formerly of Qeqertarsuaq and now of Nuuk, and their family, to this book. Their patient teaching and warm friendship contributed enormously both to this research and to the experiences my family and I enjoyed in Greenland. *Ilissinnut tamassi, qujanarsuaq.*

This book is based on my doctoral dissertation, which was completed at the University of East Anglia (UEA), School of Development Studies, United Kingdom, under the supervision of professors Piers Blaikie and Nick Abel. Professor David Seddon at UEA also provided valuable advice and guidance. The fieldwork on which it is based was approved by the Commission for Scientific Research in Greenland and the Greenland Home Rule Government. Funding was provided in part by the Wenner-Gren Foundation for Anthropological Research and by the National Science Foundation, Arctic Social Sciences Program (OPP-9112900). During the course of this research, I enjoyed opportunities to work with faculty and staff at the University of East Anglia, School of Development Studies, *Ilisimatusarfik* (the University of Greenland), the University of Copenhagen's Institute of Eskimology and Arctic Station (Qeqertarsuaq), and Scott Polar Research Institute, University of Cambridge, UK.

Many people assisted me in the preparation of this book. Ingmar Egede, Amalie Jessen, Milton M. R. Freeman, Jens Dahl, and Einar Lemche read portions of earlier drafts and provided valuable comments. However, the analysis presented here and any mistakes or omissions are mine alone. In Greenland, I am indebted for kind assistance provided by the Honorable Pâviâraq Heilmann (Minister for Fisheries, Hunting, and Agriculture), Amalie Jessen, Bjørn Rosing, H. C. Petersen, Robert Petersen, Aqqaluk Lynge, Hans Peter Grønvold, Finn Steffens, Ole Marquardt, Claus Andreasen, Per Langgård, Anton Siegstad, Hansi Kreutzmann, and Aqqaluk Rosing-Olsen. Others who have contributed directly or indirectly to this effort include Ole Bennike, Pernille Bennike, Finn Lynge, Jens Brøsted, Mads Fægteborg, Henning Thing, Morton Rasch, Alfred Jakobsen, Finn Larsen, Finn O. Kapel, Mark Nuttall, Klaus Georg Hansen, Susanne Dybbroe, the late Poul Møller, and Jens Møller.

Oran R. Young and Gail Osherenko, editors of UPNE's Arctic Visions series, encouraged me to transform a conventional dissertation into a (hopefully) more readable and engaging manuscript. Phyllis Deutsch and Mary Crittendon at UPNE guided me through processes of editing and production. Dixon Jones of the University of Alaska Fairbanks's Rasmuson Library graphics office ably produced the maps and illustrations. Sue Mitchell and Debbie Miller proofed the manuscript and provided valuable suggestions. Colleagues at the University of Alaska Fairbanks—particularly Patrick Dubbs, Raymond Barnhardt, and Bernice Joseph—provided encouragement and a pleasurable working environment for completion of the book.

Finally, I wish to acknowledge the lifetime of support and encouragement provided me by my mother, Alicelee F. Ewan, my late stepfather, Raymond J. Ewan, and my uncle, Henry P. Caulfield, Jr.

Greenlanders, Whales, and Whaling

Introduction: Whaling and Co-management in Greenland

I n 1991, while a breathless world watched the Soviet empire disintegrate and Bosnia slide into chaos, the United Nations quietly celebrated the "bloodless revolution" of 1979 that created Greenlandic Home Rule. At a major UN conference on indigenous self-governance, speakers from around the world praised Greenlanders' achievement in negotiating Home Rule from Denmark. For them, Home Rule provides a successful model for transforming relationships between indigenous peoples and former colonial powers in a world undergoing profound political and economic change. The United Nations reiterated this view in its International Decade of Indigenous People, highlighting difficulties facing Greenlanders and others situated on the world's peripheries. In doing so, it pointed to parallels between processes of self-determination in the Arctic and those of tribal peoples of Asia, Latin America, Africa, and elsewhere in the developing world.

Through Home Rule, Greenlandic Inuit people (*Kalaallit* in the West Greenlandic language) have achieved what many regard as the greatest degree of political autonomy of any indigenous peoples in the Arctic. In the process, Greenland, or *Kalaallit Nunaat*[1] (map 1), ended over 250 years of colonial control by Denmark and set the stage for a new form of indigenous self-determination, drawing upon both Inuit customs and Scandinavian social democratic traditions. Although Greenland today remains part of the Danish realm, its 55,000 people—85 percent of whom are Inuit—are undertaking a process of nation-building that enhances political autonomy and strengthens ties to the world economy.

Despite these achievements, many outside Greenland continue to think of it as a land of icebergs, seal hunters, and kayaks, far removed

MAP I. Map of Greenland (*Kalaallit Nunaat*)

from modern industrial society. As we shall see, this image is an illusion. Greenland's history is not simply one of an isolated peoples' ongoing battle against a harsh and demanding environment; it is also *our* history—a history of Euro-American expansion and colonization, and of growing integration of indigenous societies in the global economy. In Greenland today, kayaks and seal hunting co-exist side by side with ocean-going trawlers, cellular telephones, and hot dog stands. Greenlandic traditions are continually being transformed, creating new opportunities and new contradictions. Like many peoples around the globe, Greenlanders seek to blend ancient traditions with new realities in pursuing sustainable development—"development that meets the needs of the present without compromising the ability of future generations to meet their own needs."[2] Understanding Greenland's development dilemmas gives us more than just insight into how indigenous peoples are coping with a changing world; as we shall see, it also provides us with a sense of the challenges *we* face as well.

This book is about Greenlanders, whales, and whaling. It focuses on Greenlanders' hunt for minke and fin whales, which falls under the International Whaling Commission's (IWC) "aboriginal subsistence whaling" management regime. The minke whale (*Balaenoptera acutorostrata*) is a smaller baleen whale averaging about ten meters in length. The fin whale (*Balaenoptera physalus*) is one of the world's largest baleen whales, averaging about twenty-four meters. Whaling for these and other species is controversial.[3] Our knowledge of their status is growing but remains limited. The controversy over whaling, however, often has less to do with biology and resource management than it does with conflicting ideas about human-environment relations and about the very concept of sustainability.

This book explores these conflicts through an analysis of Greenland's management regime for whaling. Social scientists define a regime as a "social institution composed of agreed-upon principles, norms, rules, and decision-making procedures that govern the interactions of actors in specific issue areas."[4] Greenland's whaling management regime is particularly useful for highlighting relationships between indigenous self-determination and sustainability. In recent years, indigenous peoples around the world have begun to develop mechanisms for *co-management* of renewable resources. Co-management allows users, resource managers, and other stakeholders to share power in managing common property (or common pool) resources such as whales.[5] It can include an array of institutional innovations at various levels designed to ensure sustainable and equitable use of renewable resources. Often it involves blending Western science and indigenous knowledge in devising effective management systems.

Co-management is viewed by many as a hopeful mechanism for sustainable development and for resolving persistent resource management conflicts. I caution, however, against viewing co-management as a panacea for solving such conflicts, for it can have considerable trans-action costs for indigenous societies. These often-overlooked costs (so-cial, economic, and cultural) can result from the burdens of negotiating and sharing responsibilities with the state. They can manifest them-selves externally, in the broader management arena, but can also appear internally within indigenous societies. For example, this study reveals how interactions between hunters and distant political forces can lead to disruption and stress within indigenous societies, even within com-munities and families. Co-management can also create indigenous elites and bureaucratic structures—incipient forms of social differenti-ation—that challenge small societies already pressured by change. Al-though co-management's emphasis on power-sharing and conflict me-diation often strikes a positive chord in liberal democracies, its uncritical application could lead us to overlook important conflicts and contradictions in managing common property resources.

For indigenous societies in the Arctic, I further argue that successful co-management is not an end in itself but rather is a *process* closely linked to self-determination. Effective co-management in the North is commonly associated with broader agreements about indigenous rights.[6] Once a broad framework for addressing political and economic rights is in place, co-management provides a process and a framework for on-going discussion and conflict resolution. Absent these broader agreements, however, co-management can too easily become co-optation; a situation one indigenous leader disparagingly characterizes as "*we* cooperate and *they* manage."

Home Rule provides Greenlanders with such a framework for co-management. Understanding Greenland's whaling regime thus re-quires an awareness of what Home Rule means and how it came about. It also requires an interdisciplinary, multilevel, and historically in-formed account of both global and local dynamics; that is, we need to understand the "vertical slice" of processes, institutions, and relation-ships that extends from the household and community to the interna-tional stage. Writing such an account is no small task. In this attempt, I hope to convey both the contextual richness of contemporary Green-landic life and the profound impact of world systems on the Arctic. In my view, this multilevel approach is crucial if we are to understand the complex relationships connecting Greenland with global systems.

Inuit people and their ancestors have lived in Greenland for some four thousand years. Danish colonization in 1721 initiated Greenland's gradual penetration by the world economy. Until World War II, Greenland was largely a closed society, administered by powerful colonial authorities and dependent almost entirely on trade with Denmark.[7] World War II and the presence of American and Canadian forces changed all that. In 1953, Denmark formally ended colonial rule, paving the way for Greenlanders to become full Danish citizens and for Greenlandic-Danish relations to be "normalized."

Greenlanders, however, soon found themselves subjects of two massive, state-controlled "modernization" programs (referred to colloquially as G-50 and G-60). Anthropologist Jens Dahl argues that this postcolonial era brought to Greenlanders the worst effects of colonialism.[8] Although this argument may be an overstatement (many Greenlanders also supported these initiatives), there is no doubt that this era profoundly changed Greenland. The Danish state invested heavily in Greenland's commercial fisheries sector, importing large numbers of Danish workers to build new fish processing plants and other infrastructure. To meet increased demand for labor in processing plants, the postcolonial state used a controversial resettlement program to entice or coerce Greenlandic families into leaving smaller settlements. A number of settlements were completely depopulated, and Greenlanders were increasingly relegated to spectator status in a process of profound change.

In the 1970s many Greenlanders, including a young and vocal elite, began resisting these developments. Following a decade of agitation and turmoil, Greenlanders finally achieved Home Rule in 1979. The result was significant political autonomy within the Danish realm. Today, Greenlanders elect their own parliament (the *Landsting*) and govern nearly all matters internal to Greenlandic society. Impressive as this achievement is, however, economic autonomy has been much more elusive. Today, Greenland's dependence on the world economy is striking. A single product—deep-water shrimp (*Pandalus borealis*)—comprises over 90 percent of all exports.[9] Greenland markets shrimp throughout the world, to customers as far away as London and Tokyo. Nearly three-quarters of all Greenlanders work in the wage economy. Less than 20 percent make their livelihood predominantly from household-based hunting and fishing; furthermore, social differentiation in Greenlandic society is increasing, with some 15 percent of the population earning over 50 percent of all taxable income. The state is a major player in virtually all sectors of Greenland's economy. Its involvement is buttressed by a substantial annual subsidy of over $500 million from the Danish government, amounting to nearly half the island's gross national product.

Having recognized Greenland's vulnerability, its leadership is attempting to diversify and strength the island's economy. Home Rule officials are expanding markets for fisheries products, promoting mineral exploration, expanding tourism, privatizing public sector services, and cutting government expenditures. In the words of Greenland's economic minister:

As an integrated part of the world market we will, in the years ahead, have to adapt our economic life to an international outlook to an even greater degree. The opening of the internal market within the European Community . . . provides an unambiguous signal that a closed and protected Greenlandic economy will not be able to survive.[10]

These economic initiatives clearly further Greenland's integration with the world economy, yet, in doing so, they raise new fears about uneven development, particularly within smaller and more remote settlements. Many view these settlements as repositories of "pure" Greenlandic values.[11] These new initiatives could marginalize them even further, creating new peripheries *beyond* the periphery.

Greenland's economic future remains highly uncertain. As an increasingly integrated part of the world economy, it faces growing population pressures, increasing urbanization, and expanding social differentiation. It could also face profound changes resulting from global climate change. Given these challenges, can Home Rule match up to its promise for indigenous self-determination? Can Greenlanders achieve political autonomy without greater control over resources vital to their economy? Can a sustainable economy be built on Greenlandic terms? These questions, I suggest, are important not only for the Arctic but for peripheral societies throughout the world seeking a sustainable path toward the future.

Toward sustainability and self-determination: Greenland's co-management regime for whaling

Whales have been a part of marine-based economic strategies in Greenland for over four thousand years. As in other aspects of Greenlandic life, however, whaling has a history of ecological, economic, and political marginalization. Before Danish colonization in 1721, Greenlanders primarily caught bowhead whales (*Balaena mysticetus*) in the tradition of their Thule Inuit ancestors. Inuit spiritual beliefs and practices governed hunters' relationships with whales, and the small and dispersed Greenlandic population had little impact on whale resources. As long as hunters maintained a proper spiritual relationship with their prey, there was little need for a more formalized system of property rights.

European whalers decimated bowhead stocks in the seventeenth and eighteenth centuries, however, severely disrupting Greenlandic whaling. Danish authorities undermined Greenlandic customs and traditions, including those associated with whaling. Greenlanders lost many of the culturally based mechanisms for regulating and distributing whale catches that had served them for generations. By the early twentieth century, both Greenlandic society and bowhead stocks were at a low ebb, and Danish authorities took it upon themselves to initiate European-style whaling on behalf of Greenlanders. Using a Norwegian ship equipped with a harpoon cannon, Danes began catching whales to supply local communities with meat and *mattak* (whale skin). Even during this era, hunters in a few Greenlandic communities persisted in catching humpback (*Megaptera novaeangliae*) and other whales; a reflection of the continuing prestige associated with whaling in Greenlandic society. In the late 1940s and early 1950s, Greenlanders themselves revitalized whaling using fishing vessels and harpoon cannons. Because bowhead stocks remained low, they focused the new technology on other whale species, notably minke, fin, and humpback whales.

Ruthless exploitation of whales worldwide in the early twentieth century led Denmark to join other nations in signing the 1946 International Convention for the Regulation of Whaling (ICRW). This convention serves as the basis for international management of large whales today and enables "aborigines" to catch whales for local consumption, even where commercial whaling is prohibited.[12] As a Danish colony, Greenland was not involved in this agreement, but in later years, the aboriginal whaling provisions of the convention had a significant impact on Greenland. In 1985, the International Whaling Commission, formed under the ICRW, surprised Greenlandic hunters by reducing their allotment of minke whales by half and eliminating entirely their catch of humpback whales. By this time, the aboriginal management regime had become a major issue for whale preservation groups, which challenged Greenlandic whaling on the grounds that it could not be considered "traditional" because hunters used technology like harpoon cannons and because local markets for whale products existed. In recent years, contentious debates in the IWC have increasingly focused on the ethics of whaling, with some anti-whaling groups contesting the right of Greenlanders and other indigenous peoples to take whales at all.

Co-management of aboriginal whaling in Greenland today is influenced by a variety of different players, including the IWC, the Danish government, nongovernmental organizations, Home Rule and municipal authorities, and hunters themselves. In practical terms, co-management means that the IWC sets quotas for minke and fin whales but leaves internal regulation, monitoring, and enforcement to the Home Rule state; however, the guidelines under which Greenland is represented in international forums regarding whaling remain unclear. As it now stands, Greenland is but one part of the Danish delegation to the IWC. In nearly all other management contexts involving renewable resources (e.g., North Atlantic fisheries), Greenland acts autonomously without having to abide by a common Danish position. As we shall see, this fact creates considerable tension within the realm.

Despite these difficulties, I argue in this book that Greenland's whaling regime is increasingly effective in responding both to global concerns about the viability of whale stocks and to the needs of local communities. It incorporates indigenous knowledge and practice in management and has buffered hunters from the extraordinary political pressures surrounding whaling issues on the international level. The development of this regime also has its costs, however. Many hunters believe that the IWC is dominated by anti-whaling forces. Some advocate that Greenland withdraw from the organization, believing that hunters' interests are increasingly in conflict with those of the Danish government. Some hunters are distrustful even of Home Rule management. In a small settlement along Greenland's extensive coastline, management from Nuuk may feel little different than that from a colonial capital. Although these voices remain a minority, they illustrate the ongoing contradictions within Greenland's whaling regime and the challenges facing Greenlandic society in carrying out whaling sustainably.

Understanding co-management in the Arctic

Co-management is designed to foster more appropriate, efficient, and equitable management of renewable resources. The growing literature about co-management is part of an evolving theoretical framework within common property theory that challenges Garrett Hardin's powerful but flawed thesis about the "tragedy of the commons."[13] Hardin's essay catalyzed concerns about resource depletion resulting from common property conflicts. His thesis seemed particularly appropriate in explaining overexploitation of certain whales species in the nineteenth and early twentieth centuries.[14] His solution for these tragedies focused on expanding private property rights and developing strict systems for government control; yet Hardin's argument was flawed because he confused common property resources with open-access resources, which have no constraints on use.[15] Under open-access conditions, tragedies can and do occur, especially where there are no effective institutions to protect resources. Case studies from throughout the world, particularly those involving indigenous peoples, show, however, that community-based management can often promote resource sustainability through decentralized decisionmaking and expanded user participation.[16] Indeed, Acheson argues that

few societies or even local-level communities have no restrictions at all on the use of resources. . . . Contrary to what . . . theorists assert, privatization and government control are not the only mechanisms to affect the use of natural resources. There is a middle way: rules developed at the community level.[17]

The argument for co-management in the North is premised on two ideal types of resource management systems: an indigenous system and a state system.[18] The indigenous system, found in Inuit and other Native American societies, is based upon a collection of unwritten rules or social norms passed down through generations in the form of oral traditions. Sanctions for violation of these unwritten rules or norms are imposed through social pressure within a community of resource users. In contrast, the state system uses written law, rules, and regulations. It enforces these through a formal system of sanctions, including such things as licenses, permits, and harvest restrictions. In the North, where distances are vast and enforcement of regulations is difficult, implementing the state system has been problematic. Many observers believe that the two systems must be integrated for the social and economic health of indigenous communities and to achieve environmentally sustainable and culturally appropriate development.[19]

Co-management regimes can have varying levels of power-sharing between users and the state.[20] Research has shown that there can be a continuum of these arrangements, ranging from simple information-sharing to a "partnership of equals," with full community control. Berkes describes a well-functioning regime as one with:

- a minimum (or absence) of disputes and limited effort necessary to maintain compliance; the regime will be *efficient*
- a capacity to cope with progressive changes through adaptation, such as the arrival of new production techniques; the regime will be *stable*
- a capacity to accommodate surprise or sudden shocks; the regime will be *resilient*
- a shared perception of fairness among the members with respect to inputs and outputs; the regime will be *equitable*.[21]

In this study of co-management and self-determination in Greenland, I make three main arguments. First, I emphasize that ideological conflicts in the IWC about whaling are rooted in idea-systems born of differing modes of production in Greenlandic and Euro-American societies. Greenlanders' mode of production is largely kin-ordered and based upon the concept of mutual security. In contrast, the dominant Euro-American mode of production is capitalist and emphasizes individualism and capital accumulation. These differing modes give rise to conflicting idea-systems about environment, development, and change in the Arctic. Within the IWC, they underlie contentious and often bitter debates about the future of indigenous whaling.

Second, I argue that Home Rule provides Greenlanders with a vehicle for meaningful participation in the international whaling regime. Because of Home Rule, Greenlanders are not simply passive victims in IWC debates, as they might be if the country was simply a Danish territory. They are able to interact effectively with other stakeholders involved with whaling issues and to restructure relationships with powerful macro-level forces. They do so despite the limitations of being represented by a common Danish IWC delegation. The best evidence of this autonomy is the Home Rule government's decision in 1992 to join Iceland, Norway, and the Faroe Islands in creating a North Atlantic Marine Mammal Commission (NAMMCO). NAMMCO is a regional institution for conservation, management, and study of marine mammals, and it potentially could be an alternative to the IWC.[22] As we shall see, Greenland's participation in NAMMCO brings both risks and opportunities.

Finally, I emphasize that co-management is no panacea for effective resource management in the Arctic. The expanding literature about co-management highlights its successes but rarely addresses its costs and the internal dynamics generated by participation. This case study from

Greenland shows how power relationships within and between extended families and communities are altered by participation, and how local decisionmaking patterns are changed. It reveals the costs and challenges of participating in co-management for indigenous political institutions like the Home Rule state.

Today, Greenland's whaling regime is at a formative stage. The Home Rule government is moving deliberately but cautiously in implementing new regulations, mindful that the regime requires political acceptance from hunters themselves. At the community level, regulations and quotas have already altered the social relations of whaling, changed decisionmaking structures, and contributed to some degree of social differentiation. New quotas and regulations create political tensions within local municipalities, making enforcement of new regulations a sensitive and sometimes difficult matter. Additionally, recent developments in the IWC—notably the creation of an Antarctic whale sanctuary—are generating new tensions between hunters, the Home Rule government, and Danish authorities. Having said this, however, Greenland's whaling regime appears to have increasing legitimacy in the eyes of most hunters.

In her book *Co-operative Management of Local Fisheries*, Evelyn Pinkerton underscores the importance of a regime's internal dynamics and suggests that co-management research focus on them.[23] In particular, she calls for research that tests hypotheses about relations between peripheral resource communities and the state, relations among more and less powerful communities, and relations among more or less powerful or wealthy individuals within one community. This book addresses precisely these issues. As I argue in later chapters, internal dynamics can influence the character and pace of indigenous responses to crises in co-management. They can also place significant constraints on the ability of emerging indigenous political institutions like Home Rule to manage resource uses. Taken together, these dynamics can strongly influence the creation of effective, sustainable development strategies in Arctic societies.

This book, moreover, addresses one of the major questions posed about development and change in the Arctic; that is, to what extent can indigenous societies on the world's periphery shape their interactions with larger political and economic forces? Today, most acknowledge that the recent history of the Inuit people is one of dependency and internal colonialism, where they have been marginalized within larger nation-states.[24] Too often, however, efforts to apply dependency theory to the Arctic have led to sweeping assertions about all-powerful centers and passive peripheries;[25] situations where, in the words of anthropologist Harvey Feit, "micro-level changes originate primarily externally, and more particularly that local-level responses are simply reactive, the local population having neither the power nor the means to generate unique or effective responses."[26] In this study, I argue against assumptions that "modernization" is inevitable or that local societies are powerless to shape events. Although the Arctic may be an underdeveloped periphery, it is far from passive. In chapter 5, I focus in particular on mechanisms of resistance and countervailing force—what anthropologist James Scott calls "weapons of the weak"—used by Greenlanders to counter external forces challenging whaling.[27] These mechanisms illustrate how societies on the periphery do indeed have tools at their command that can be used for empowerment and self-determination. As Feit points out,

[W]hen hunting peoples in developed states are able to mobilize some political/economic leverage in the macro-arena then it may be possible for them not only to resist external pressures leading to a restructuring of their own social fabric, but they may also be able to restructure the relationship between themselves and the impinging institutions. The extent of such a restructuring is variable, and the means of reorganizing relationships between a hunting society and macro-societies necessarily involve the *creation and introduction of new institutions*. (emphasis added)[28]

Feit's focus on appropriate institutional development is based on the idea that the developed center may not be "all-powerful and hyper-integrated."[29] Instead, the relationship may be one of interdependency, where peripheral societies marshal economic, political, and cultural resources to reshape interactions between center and Arctic periphery.[30]

To test this empirically, Feit calls for studies that encompass the broad range of interactions between centers and peripheries:

We might say that we need to move from the single-focus study of dependence, to a wider framework that, without abandoning such study, also includes study of the means of action by which autonomy may be created among the constraints causing dependence. . . . This will have to involve, in my view, analyses of local beliefs and knowledge, of possible sources of local power, and of the effectiveness of decisions and actions at all levels. This will require the linking together of analyses of both the logic and the incoherence of macro-level structures with comprehensive decision-making analyses of the micro-level knowledge, power, imagination and initiative.[31]

This case study of global and local dynamics in Greenland tests Feit's hypothesis. If we are to discover new ways of promoting sustainability in Arctic communities, we must begin by understanding how indige-nous peoples can create new institutions for self-governance that strengthen control over resources vital to their livelihood. Put in an-other way, research on development and change in the Arctic must "take cues from societies whose very existence 'development' has al-ways threatened."[32] This approach not only has theoretical importance but has practical significance as well. In Feit's words,

it is not simply a question of the transformation of hunters and gatherers into something else: farmers, pastoralists, slum dwellers, ethnic minorities, prole-tarians, specialized laborers, or welfare recipients. It is also a question of the transformation of hunting societies into new and potentially viable forms of hunting societies, with diverse productive organizations, consumer goods, complex imported technologies, and extensive state intervention and relation-ships.[33]

Chapter organization

This study draws upon theoretical perspectives and empirical data from a number of fields, ranging from political economy and economic an-thropology to marine biology and policy studies. In my view, interdisci-plinary studies of this type are essential for understanding sustainability and self-determination in the Arctic. However, interdisciplinary re-search has its limitations. As Marshall Sahlins points out, it can be char-acterized as "the process by which the unknowns of one's own subject are multiplied by the uncertainties of some other science."[34] I quite agree with Sahlins, but also with Cronon's response: "Like Sahlins, I think the benefits of interdisciplinary work outweigh the dangers, but I share his sense of risk."

The book is divided into two parts. Part One focuses on longstand-ing connections between Greenland and global political and economic systems; on Greenland's history as a Danish colony and—more re-cently—the development of Home Rule and an economy highly de-pendent on world markets and Danish subsidies. I illustrate these dy-namic relationships at the local level through a case study of production relations from Qeqertarsuaq (in Danish: Godhavn) Municipality, lo-cated in the Disko Bay region of northern West Greenland. Part Two links local whaling practices with the global whaling management re-gime. It again draws upon data from Qeqertarsuaq to illustrate the two major types of whaling today in Greenland: fishing vessel and collective whaling. It describes the development of co-management for whaling under Home Rule, including conflicts within local communities, within the Danish realm, and within the IWC itself. The concluding chapters examine Home Rule initiatives to develop more appropriate institutions for managing marine mammals—notably through NAMMCO (North Atlantic Marine Mammal Commission)—and review the prospects for sustainable use of marine mammals in Greenlandic society in the years ahead.

PART ONE

CONNECTIONS—GREENLAND AND WORLD SYSTEMS

. . . the world of humankind constitutes a manifold, a totality of interconnected processes, and inquiries that disassemble this totality into bits and then fail to reassemble it falsify reality. Concepts like "nation," "society," and "culture" name bits and threaten to turn names into things. Only by understanding these names as bundles of relationships, and by placing them back into the field from which they were abstracted, can we hope to avoid misleading inferences and increase our share of understanding.[1]

History and Political Economy
in Greenland

Introduction

Histroy is vital to understanding relationships between Greenland and global political and economic forces. Greenlandic and Euro-American histories first converged in the thirteenth century with encounters between Inuit hunters and Norse colonists along Greenland's western shore. This contact may well have occurred about A.D. 1266, when a party of Norsemen encountered signs of indigenous Greenlanders in Disko Bay.[2] Historian Gwyn Jones describes these first interactions as the culmination of a global process that was "at least 12,000 years and some 500 generations in the making."[3] In the North Atlantic region, he notes, the "two human populations expanding around the globe in opposite directions met for the first time."

In the seventeenth and eighteenth centuries, European expeditions to Greenland brought back wondrous images of the people who lived there. They even brought back indigenous captives clad in sealskins and furs, stolen from their families to become curiosities in Europe. These captives provided Europeans with their first real impressions of the Inuit: seemingly a "primitive" but hardy people, ever cheerful in the face of an unforgiving environment.

Though these events are long past, the images and stereotypes of that time continue to influence our perceptions of Greenland and the Arctic today. As anthropologist Hugh Brody observes, the Arctic in Euro-American consciousness is

identified with Nature at its most natural. The Arctic exists in the popular consciousness with much the same efficacy as the Sahara or the Orient . . . a place of unremitting discomfort, where the raw ingredients of Nature are undiluted by those more balmy elements in which civilization is thought to thrive.

Its seas are frozen and its lands are barren. It is a place where society is not possible, and individuals manage to struggle alone against overwhelming odds, wasted physically and spiritually by the elements. . . . The Arctic is popularly thought to be implacably antagonistic to Culture.[4]

Greenlanders, as subjects of these early accounts from European exploration, continue to live with this legacy—the image of "primitives" living in simple societies in an environment viewed by outsiders as utterly inhospitable. As Stefansson once wrote:

They gathered their food with the weapons of the men of the Stone Age, they thought their simple, primitive thoughts and lived their insecure and tense lives. . . . I had nothing to imagine; I had merely to look and listen; for here were not remains of the Stone Age, but the Stone Age itself.[5]

In the age of Arctic supertankers, land claims, and eco-tourism, we've clearly moved beyond these crude stereotypes, yet their legacy persists to a remarkable degree. Most North Americans and Europeans today would be hard pressed to provide more than simple impressions of the Arctic as a place of perpetual cold inhabited by "Eskimos," though some might also view it as a rich storehouse of oil and minerals.

Perhaps in reaction to earlier stereotypes, we now encounter a new image of the Arctic—that of Inuit as lay ecologists or "natural conservationists." As Euro-Americans become more environmentally aware, some characterize indigenous societies as adapting perfectly to ecological conditions, unable to err in interactions with a harsh environment. This perspective is "practically a cliché in the literature concerned with preindustrial peoples," writes one observer:[6] the notion of "noble savage" living on in a changing world. This shift in our perceptions of Inuit and other foraging societies is telling. As Bender and Morris point out,

the way in which perceptions of gatherer-hunters have swung—still swing—in the last hundred years, between "not far above the anthropoid apes" . . . and (successful) "primitive communism" . . . , suggests that we use them as a foil for our own societies. The length of time it has taken for the notion of "pristine" present-day gatherer-hunters to be abandoned suggests that being "pristine" was somehow important to us: we did not want to know what they were like after we had decimated, demoralized and destabilized them; we wanted to know what they were like beforehand, because we wanted to know what *we* were like.[7]

These lingering perceptions reveal problems in our understanding about how Greenlanders and other northern peoples are changing through interaction with world political and economic systems. As Bird-David notes,

if to regard the hunter-gatherer social system as generated by a foraging economy can be described as too "isolationist," then the full swing to presenting
it as the outcome of trade-contact with adjacent societies can be seen as too
"integrationist." Both explanations can be criticized as partial and simplistic. . . . The key point is that both explanations are anachronistic, in the sense
of still being framed within the paradigm of the isolated and self-sufficient
hunting and gathering society.[8]

In his book *Europe and the People Without History*, anthropologist Eric
Wolf argues that social sciences generally—and anthropology in particular—too often perpetuate the idea of foraging societies like that in
Greenland as anthropological isolates; bounded communities devoid
of connections to broader political and economic systems.[9] According
to Wolf,

the concept of the autonomous, self-regulating and self-justifying society and
culture has trapped anthropology inside the bounds of its own definitions. . . .
[T]he compass of observation and thought [in anthropology and other social
sciences] has narrowed, while outside the inhabitants of the world are increasingly caught up in continent-wide and global change.[10]

In the years since Wolf made this charge, it has become almost fash-
ionable for anthropologists and ethnographers to emphasize that indigenous peoples have their own histories. Most mainstream anthropologists today would agree with Kuper that the notion of primitive
society—society lacking a history—is an illusion.[11] Indeed, Greenlanders and other Inuit are perhaps the best known of all indigenous
societies studied by anthropologists.[12] Far from living in isolation, they
have a history of interaction with Euro-Americans that spans at least
eight hundred years. Even today, however, our understanding of
Greenlandic history comes largely from accounts of early whalers, missionaries, and colonial administrators; accounts that "flash-freeze" the
technologies, practices, and belief systems of the early contact period
as a benchmark for evaluating everything that follows.

Understanding Euro-American perceptions of Greenlandic society
is more than an academic exercise. These perceptions have a real impact on Greenlanders' livelihoods today, particularly with regard to
whaling. When Greenlanders don't fulfill *our* expectations of the attri-
butes of a lay ecologist or natural conservationist, we can feel disappointed or even cheated. When they use a high-powered rifle or an
exploding harpoon to catch a whale, we wonder why they can't be doing
it the more "traditional" way. As a leader of one anti-whaling group in
Europe put it, "if they want to kill whales in the traditional way, that's
fine by us, if nothing else about their way of life, significantly anyway,

has changed."[13] While our own world is being profoundly changed by genetic engineering, environmental pollution, and epidemics like AIDS, we seek reassurance that it is possible to live more "in harmony" with the natural world; and if Greenlanders or other Inuit don't fulfill our expectations, we can be disappointed indeed.

In this book, I argue that confrontations over Greenlandic whaling reflect profound differences in idea-systems about environment, development, and change. These idea-systems are rooted in differing modes of production. Eric Wolf characterizes a mode of production as "a specific, historically occurring set of social relations through which labor is deployed to wrest energy from nature by means of tools, skills, organization, and knowledge."[14] In Greenland, the dominant mode can be characterized as what Wolf calls kin-ordered, where kinship, sharing, and mutual security are paramount. In contrast, a capitalist mode of production is dominant in Euro-American societies, a mode having a particular emphasis on individualism and capital accumulation.

In the IWC context, conflicts over hunting technologies, humane killing of whales, sale of whale products for cash, and the contribution of indigenous knowledge to management are all informed by idea-systems born of these differing modes of production. As we shall see, in the end these confrontations over ideas usually have to do with power and control. Whose ideas will prevail becomes the main question. Wolf highlights this complex and dynamic relationship between idea-systems and political power:

[T]here is . . . an economic and political side to the formation of idea-systems, and idea-systems, once produced, become weapons in the clash of social interests. Sets of ideas and particular group interests, however, do not exist in mechanical one-to-one relationships. If a mode of production gives rise to idea-systems, these are multiple and often contradictory. They form an "ecology" of collective representations, and the construction of ideology takes place within a field of ideological options in which groups delineate their positions in a complex process of selection among alternatives. This process of inclusion and exclusion is not only cognitive; it also involves the exercise of power.[15]

History is central to this process of inclusion and exclusion, because interpretations of history differ significantly between Euro-American and Greenlandic societies. We thus need to examine Greenland's whaling regime not in isolation but through a careful historical analysis of dynamic relationships between two distinctly different societies. We moreover must understand the characteristics of their conflicting idea-systems about whaling and the distinctive modes of production that give rise to them.

Conceptual tools: World-systems analysis and modes of production

The conceptual framework we choose for understanding these conflicts is of the utmost importance. In the effort here, I draw upon Immanuel Wallerstein's world-systems analysis as a means for understanding historical development processes in Greenland and their significance to the whaling debate.[16] World-systems analysis embraces political economy's traditional emphasis on core/periphery relations, the role of the state, and questions about who bears the costs of development.[17] According to Wallerstein, it has three defining characteristics.[18] First, it uses the "world system" as the most appropriate unit of analysis for studies of social behavior; that is, it highlights dynamics between local, national, and international realms and denies that the "nation-state" can be viewed as autonomous. Second, it emphasizes that world systems are explicitly historical and dynamic; they have "beginnings, lives, and ends."[19] As such, it highlights the transitory nature of institutional structures and the fact that there are uncertain transitional periods between one historical system and another. The third characteristic is a particular view of the world system in which we live today, the capitalist world economy. Wallerstein views this capitalist world economy as having:

1. the ceaseless accumulation of capital as its driving force,
2. a structural division of labor between core and peripheral zones,
3. a history where hegemonic states (largely in Europe) have expanded their influence to incorporate the entire globe,
4. processes whereby antisystemic movements simultaneously undermine and reinforce the system, and
5. cyclical rhythms and secular trends that contribute to inherent contradictions of the system.

Significantly, world-systems analysis also recognizes variations within core and peripheral zones, opening up the possibility that societies on the margin of the world economy can shape their future, at least to some degree.

World-systems analysis highlights the *social construction* of human-environment relations. This approach helps us steer clear of the pitfalls of environmental determinism and to see that environmental problems arise not simply from misperceptions of some normative reality revealed by Western science but from "the clash of plural realities, each using impeccable logic to derive different conclusions . . . from different premises."[20] It also emphasizes the "fundamental importance of racism and sexism as organizing principles" in the capitalist world economy.[21]

World-systems analysis provides a framework for moving beyond the tired clichés and tortured discourse about development so prominent in the past. As Michael Redclift points out:

> The development discourse is usually conducted through comparing the claims of neoclassical economics and Marxist political economy. However, both approaches have been found wanting, notably in their inability to provide an alternative to industrial society. The growth of interest in our responsibilities to nature, in alternatives to alienated labor and commodity fetishism, and the attention which feminists have paid to the social construction of gender, should give us cause to reflect on the trajectory which "development" has taken in industrial society.[22]

One of the particular weaknesses of development discourse has been its inability to capture fully the dynamics of non-European societies, like that in Greenland. At issue is the framework we use for understanding changing relationships between societies on the periphery and those at the center. Are we to assume, as Murphy and Steward once did, that "the aboriginal culture is destined to be replaced by a new type which reaches its culmination when the responsible processes have run their course"—the result being a loss of self-determination and autonomy?[23] In Greenland, can we expect that indigenous economies will inevitably and inexorably be absorbed by the world economy?

The concept of a mode of production is helpful here because it focuses attention on the underlying framework for social organization of labor. The concept originated with Marx, who viewed it as an essential starting point for analysis.[24] Marx, however, used the concept ambiguously when it came to non-European societies, referring variously to primitive, communal, ancient, slaveholding, feudal, Asiatic, and peasant modes. Theorists in the twentieth century have continued to struggle with applying it to peripheral societies.[25]

Wolf provides a way out of this confusion by refocusing our attention on the utility of the concept, not as a box for categorizing societies, but as a means for understanding their underlying dynamics:

> [I]t is immaterial whether Marx was right or wrong—whether he should have postulated two or eight or fifteen modes of production, or whether other modes should be substituted for those suggested by him. The utility of the concept does not lie in *classification* but in its capacity to underline the strategic relationships involved in the deployment of social labor by organized human pluralities.[26]

Wolf goes on to offer a parsimonious yet elegant set of categories for understanding how social labor is organized for production: a capitalist mode, a tributary mode, and a kin-ordered mode. The latter, like Sahlins's domestic mode of production, is based largely on kinship and on production for use rather than for exchange.[27] Wolf acknowledges that other modes could exist as well, but that, in any case, these constructs

should not be taken as schemes for pigeonholing societies. The two concepts—mode of production and society—pertain to different levels of abstraction. The concept of society takes its departure from real or imputed interactions among people. The concept of mode of production aims, rather, at revealing the political-economic relationships that underlie, orient, and constrain interaction.[28]

If a society like that in Greenland is characterized as kin-ordered, does that mean it has no relationships with the world economy? Of course not, because all human societies are influenced to some degree by global forces. Wolf's approach, however, distinguishes between societies with a capitalist mode of production and those that are kin-ordered yet have links to world systems through exchange relations. His approach explicitly acknowledges the diversity of relationships within the world economy:

The capitalist mode of production may be dominant within the system of capitalist market relations, but it does not transform all the peoples of the world into industrial producers of surplus value.... [This approach also] allows us to take note of the heterogeneity of the different societies and sub-societies making up the system rather than obliterating that heterogeneity in dichotomies such as "core-periphery" or "metropolis-satellite."[29]

In the Greenlandic context, this recognition also allows us to move beyond simplistic assumptions of a passive and powerless periphery to explore the complex and dynamic history of interrelationships between Greenlanders and world systems. In later chapters, we return to this discussion about modes of production in examining the significance of Greenlandic whaling at the community and household level.

Distinctive characteristics of Greenland's history

Greenland's history has been profoundly influenced by dynamic ecological conditions. Greenland straddles a boundary between two major climatic systems: North Atlantic maritime and polar arctic. Its enormous icecap covers over 1.8 million square kilometers, or nearly 90 percent of the island's land mass. The ice exerts a marked influence not only on local microclimates but on the climate of the entire North Atlantic region. Mean temperatures in Greenland are below +10° C, even in the warmest months, and extremes reach −30° C or more in winter.

Figure 1 shows the relationship of climate change to patterns of human habitation over time in Greenland. The arrival of Thule Inuit people and Norse colonists about the end of the first millennia A.D. coincided with a period of relative warmth. Temperatures, however, dropped dramatically in later centuries, reaching a low point between A.D. 1300 and 1500. During this period, Norse colonies in Greenland disappeared, whereas Thule Inuit society thrived. Scholars studying the Norse believe that changing climatic conditions made it increasingly difficult for them to perpetuate livelihoods more suited to conditions in northern Europe;[31] furthermore, expanding sea ice probably made it more difficult to trade for essential goods with Iceland and Norway.

These climate shifts occur frequently in Greenlandic history. In the early twentieth century, increasing air temperatures and the arrival of warm ocean currents off West Greenland created a new fishing economy based upon Atlantic cod. In recent years, however, cod catches have declined dramatically, forcing closures of processing plants built only a few years ago. Greenlanders today also face a new concern: that of potential impacts from global warming. With its enormous icecap, Greenland could face profound changes with even slight increases in global temperatures.

A second factor affecting Greenland's history is the primacy of marine resources—fish, seals, whales, seabirds—in local economies. "We don't talk much about *land* use in Greenland," commented one fisherman, "we mostly use the sea." Archaeological excavations dating back 4500 years in Disko Bay revealed that 60 percent of bones found in middens were from marine mammals, and nearly all others were from marine birds.[32] These patterns continue today, with marine resources providing the bulk of Greenlanders' food and the basis for economic survival.

Ecological uncertainty is a third factor affecting Greenlandic history. Living resources—both animal and plant—are subject to marked fluctuations. For example, cyclic changes in cod and caribou populations in West Greenland seem to occur over a seventy- to one-hundred-year period.[33] Arctic resources are further characterized by low growth rates. Once depleted, they can take a long time to recover.

These ecological factors require Greenlanders to employ a flexible,

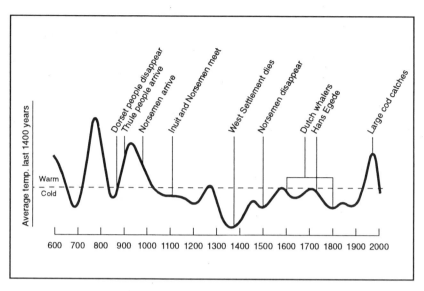

FIGURE 1. Major temperature variations in Greenland, A.D. 600 to present[30]

multispecies approach to resource use. Indeed, this flexibility may well be the key to Inuit survival in the Arctic over time.[34] When a valued resource is not available due to environmental changes, Greenlanders shift to other resources. As we shall see, however, the Greenlanders' growing interaction with the world economy is also bringing new constraints on their ability to shift easily to alternate resources.

A fourth factor affecting Greenland's history is the island's enormous regional variation in physiography, climate, and the availability of living resources. Greenland extends some 2700 kilometers from north to south, and its coastline contains a multitude of islands, bays, and deep fjords. South Greenland (*Kujataa*) lies at a latitude of about sixty degrees north, about the same latitude as Oslo in Norway or the Kenai Peninsula in Alaska. It lies well south of the Arctic Circle and thus does not have periods of winter darkness. In most areas, the ocean is free of ice year-round, making navigation and winter fisheries possible. In contrast, Greenland's northernmost point is at eighty-three degrees north; no land lies closer to the North Pole. Greenland's northern- and easternmost regions abut the Arctic Ocean and have a dry, cold climate. The ocean is ice-covered for most of the year, and people there experience long periods of winter darkness (in Qaanaaq, from November to mid-February).

These regional variations are reflected in the distinctive social adaptations of three subsocieties in Greenland: the *Inughuit* of North Greenland, the *Iit* of East Greenland, and the *Kitaamiut* of West Greenland. Though collectively they are referred to as *Kalaallit* ("Greenlanders"), each subsociety speaks its own dialect of *Kalaallisut*, the Greenlandic Inuit language; the dialects are for the most part mutually unintelligible. As we shall see, these subsocieties also have distinctive histories and economies.

Greenland to the seventeenth century: Early interactions

Ancient peoples first inhabited Greenland more than four thousand years ago, migrating east across the Canadian archipelago from Alaska's Bering Straits region. Independence I peoples were the first to reach Greenland about 2400 B.C. They were primarily hunters of muskoxen and other land animals.[35] In subsequent eras, Saqqaq (ca. 2000–1500 B.C.), Independence II (1400–600 B.C.), and Dorset peoples (ca. 500 B.C.–A.D. 1500) made their way to Greenland over the same route.

Thule Inuit, ancestors of contemporary Greenlanders, arrived in Greenland sometime between A.D. 1050 and 1100. Thule peoples excelled at seal hunting and bowhead whaling. They had a remarkable array of tools, including sinew-backed bows, kayaks, skinboats (*umiaq*; plural, *umiat*) for whaling, and dog sledges. Initially, they remained in Greenland's northernmost regions, but with climatic cooling in the twelfth and thirteenth centuries, they began to move south along Greenland's west coast.

This migration coincided roughly with the arrival of Norse colonists in south Greenland, led by Icelander Eric the Red. He explored the region in A.D. 982 and returned again in A.D. 986 with fourteen heavily loaded ships to establish two settlements: the East Settlement near Cape Farewell and the West Settlement near present-day Nuuk. From here, Norse adventurers undertook voyages to North America, settling briefly in Newfoundland and probably visiting Labrador and Baffin Island. The Norse brought with them European knowledge about farming and raising sheep, but they supplemented these livelihoods with hunting. They regularly undertook journeys northward (*Nordresetur*) to the Disko Bay region in search of walrus and narwhal ivory, walrus hides, and polar bear skins that could be traded in Europe. At the height of colonization, some three thousand Norse lived in Greenland. In the early years, they established churches, supported a bishop, and paid annual tithes both to the Norwegian king and to church authorities in Rome.

Archaeological evidence of early interaction between Thule Inuit and the Norse is scanty and inconsistent, but the available evidence suggests they were involved in sporadic trading.[36] Both may well have benefited from this contact. Inuit had ivory, much in demand in Europe, whereas the Norse had metal tools and other goods. By the fourteenth century, Inuit had moved down Greenland's west coast to the vicinity of Norse settlements. Oral traditions suggest that the two societies lived peaceably in the same area for some time, although some scholars suggest that Inuit attacks may have contributed to the Norse settlers' demise in the fifteenth century. Most agree with McGhee, however, who states that the disappearance of the Norse was probably due to a number of factors:

[Norse] decline and eventual disappearance probably had more to do with a deteriorating climate, combined with a rapid decline in the value of their commercial products as furs and ivory began to reach Europe with the growth of trade in the east and Portuguese exploration in Africa. [Norse] Greenlandic life must have stayed relatively constant, while Europe underwent the immense social changes from feudalism to mercantile capitalism. Abandoned by their Norse king and their Roman church, neither of whom any longer bothered to send ships to Greenland, and possibly harassed by European pirates, the Norse colonies were more likely the victims of economic forces than of native attacks.[37]

Climatic conditions that caused the demise of Norse settlements were ideally suited for Inuit livelihoods. By 1450 Thule people had rounded Cape Farewell and reached the Ammassalik district in East Greenland. As Greenlanders expanded their hunting areas in the sixteenth and seventeenth centuries, a robust indigenous exchange economy developed. Family groups up and down the coast left their winter homes in protected fjords and spent the summer at well-established trading sites known as *aasiviit* (singular, *aasivik*). As Petersen and colleagues note:

Since old times Greenlanders from different areas met in certain places where possibilities for seasonal hunting were optimal. These places also had a function as markets. Baleen and fishing lines made of baleen were important barter objects on these occasions, because the catch of great whales mainly took place in certain areas. . . . Food products from whales were apparently not bartered in the same manner, but dried meat, muktuk [*mattak*] and flippers were often used when entertaining visitors from other settlements.[38]

At *aasiviit*, people from southern Greenland could obtain valuable resources, like soapstone, whale baleen, and driftwood from those living
to the north. *Aasiviit* were also important social gatherings, where one
could find a spouse, consult with a shaman from far away, or visit with
distant family members.[39]

Greenlanders today view this period prior to Danish colonization as
one of political independence, where interaction with Europeans was
sporadic and had little impact on Greenlandic livelihoods.[40] By the
early seventeenth century, however, this pattern was beginning to
change in West Greenland. The rise of merchant capital in Europe,
advances in ocean navigation, and the lure of riches in polar seas soon
brought a new age of European exploration and colonialism.

Whalers, traders, and missionaries, 1721 to circa 1900

Englishman Martin Frobisher's "re-discovery" of Greenland in
1576–78 ignited European interest in Greenland.[41] The scramble in
Europe to find the fabled Northwest Passage to markets in India and
China set in motion intense competition between colonial powers.
Frobisher's voyage and that of John Davis in 1585–87 set the stage for
later voyages by Danish-Norwegian, Dutch, Hamburg, and English
merchant-capitalists.

These expeditions followed in the wake of Basque sailors, who had
been whaling in the North Atlantic as early as the eleventh century;
however, Basque whaling was undermined by vigorous competition
from the Dutch and English. In 1614, merchants in Amsterdam
founded the *Noordse Compagnie*, forming the basis for Dutch whaling
and trading in Greenlandic waters. The primary objective was to obtain
train oil, derived from whale blubber, which was used to light the lamps
of Europe. A single bowhead whale could produce as much as twenty
tons of oil.[42] Whalers initially hunted bowhead stocks near Svalbard
and along Greenland's east coast but then moved toward Davis Strait
as those stocks became depleted.

The Danish Crown, concerned about reinforcing its sovereignty,
sent out an expedition to Greenland in 1605 and also approved formation of several whaling companies. Between 1630 and 1660 other Danish and Dutch companies sent expeditions to Greenland. During one
voyage, a Dutch captain brought four Greenlanders back to Denmark,

who provided a wealth of information about seventeenth-century Greenlandic life. Unfortunately, they died of hunger and disease in Copenhagen in 1659 during Denmark's war with Sweden.

Despite setbacks, the Dutch carried on profitable trade with Greenlanders in the seventeenth century. Greenlanders preferred goods like shirts, gloves, knives, fishhooks, and specially developed replicas of the Inuit women's knife, the *ulu*. The Dutch avoided selling firearms, gunpowder, and bullets, fearing for their own safety. Typically, Dutch ships would anchor near present-day Sisimiut to trade for hides and blubber; however, as Gad notes,

the trade seems to have been rather one-sided. Except for the beads, knives, fishhooks, sword blades, ulus, and some of the tools, the Greenlanders apparently had little need for the merchandise offered; their wants were evidently quickly satisfied. But the Dutch had such a great appetite for hides and blubber that they forced their merchandise on the Greenlanders.[43]

While the Dutch prospered, a young Norwegian priest named Hans Egede was busy laying plans to bring the Gospel back to the "lost" Norse settlers. Born in 1686, Egede learned about Greenland from his father-in-law, a ship's officer. Egede thought it a travesty that the Norse had been abandoned by their church and pleaded with the Danish king for support. His initial proposals for recolonization were rebuffed, however, in 1721, the king approved the plan and established a trading company in Bergen to support the mission. Egede unsuccessfully sought a Crown-sanctioned trade monopoly, but the king, concerned about the Dutch, didn't want to provoke a confrontation.

On July 3, 1721, Hans Egede and his family arrived in Greenland and established a small settlement at Kangeq, near present-day Nuuk.[44] This action caught the attention of Greenlanders, who were used to Europeans trading but not overwintering. Local *angakkut*, or shamans, opposed Egede's religious mission;[45] this opposition proved strong enough that several Greenlandic converts were killed by their kin. Egede's mission immediately experienced economic problems. The Bergen company collapsed in debt in 1727, and from then on trade with Greenland was handled through a succession of enterprises based in Copenhagen. In 1728, Egede's colony moved to Godthåb (Good Hope), Greenland's present-day capital at Nuuk. The Crown nearly gave up entirely on the venture in 1730, but Egede convinced the king to continue. Egede expanded his mission up and down the West Greenland coastline, seeking to thwart Dutch trade by establishing new settlements. In the 1740s, the trading company initiated whaling, but in the end it, too, proved unprofitable.

In 1776, Egede's dream of a trade monopoly was achieved when the Crown closed the entire coast to foreign ships. Importantly, after 1774 all trade with Denmark was channeled through the newly created Royal Greenland Trade Department (*Den Kongelige Grønlandske Handel*, or KGH). In subsequent years, this state monopoly would have a profound impact on the development of Greenlandic society.

For Greenlanders, European colonization had devastating impact. In 1733–34, the first of several smallpox epidemics hit Greenland, and nearly half of the Greenlandic population died as a result.[46] Greenlanders also suffered from growing feuds between the mission and the trade company. Missionaries encouraged many local people to live near settlements and to become Christians, whereas traders preferred them to continue hunting away from the settlements. In some cases, Greenlanders dependent upon the stations found themselves without sufficient supplies for winter.

In an effort to cope with these problems, the Crown issued the Instructions of 1782, establishing the basis for interaction between Greenlanders and Europeans for the next one hundred years. Under the new decree, West Greenland was divided into two inspectorates; Qeqertarsuaq (Godhavn) became the administrative seat for the north, and Nuuk, for the south. The Instructions provided support for the growing number of mixed Greenlandic-Danish children. In later years, these people would form the basis for the so-called "great Greenlandic families"—catechists, trade managers, interpreters, and skilled tradespeople who are at the forefront of emerging social classes in Greenland. According to Kleivan:

Only the population mixture emanating from marriages between indigenous women and Danish or Norwegian men employed in the colonial service resulted in social processes of historic significance. First and foremost these marriages gave rise to social differentiation that has not had many parallels in the colonial history of other parts of the Eskimo world. . . . They played a most significant role in the dissemination of the new culture that emerged in the early colonial period as the product of the encounter of European and Greenlandic Eskimo culture.[47]

Despite the decree banning foreign trade, English and Dutch whalers continued to visit Greenland in the late 1700s. The Napoleonic War of 1807–1814, however, disrupted all European trade, forcing the mission to cut back its activities. A shortage of trade goods occurred, revealing the extent of Greenlanders' reliance on European goods. In 1835, the Crown instituted a new policy for Greenland, requiring that all activities in the colony pay for themselves. The government closed Greenland to outsiders, seeking to protect local people from the worst effects of Euro-American society. At the same time, the Crown introduced Greenlanders to cash. For the first time, KGH paid cash for seal-skins and other products instead of using a credit-barter system. As Petersen notes:

> This was an historic event. From ancient times, Greenlanders had to make sure that they had provisions for the coming winter. All kinds of meat and fish were dried and put up for winter food. But the implementation of the cash economy began to create a whole new relationship. . . . People developed new consumer tastes. They didn't think of money as something that corresponded to a winter's provisions. On the contrary, they developed the impression that money was simply to use here and now.[48]

The change had a profound effect, especially in central West Greenland where contact was extensive. It contributed to a pattern of what Gad characterizes as "poverty, shabbiness, and apparent hopelessness."[49] Greenlanders increasingly lost self-respect, caught between the influences of European contact on the one hand and a serious erosion of Inuit beliefs and practices on the other.

Heinrich Rink, a Danish administrator who came to Greenland in 1848, recognized this social disintegration and began working with influential Greenlanders to improve the situation. Rink encouraged publication of Greenland's first newspaper, *Atuagagdliutit*, in 1861.[50] Published using a newly developed West Greenlandic orthography, the newspaper proved to be a significant force in developing literacy and an incipient Greenlandic national identity. Rink was also instrumental in creating *forstanderskaber*, or municipal councils, for local governance. Their major purpose was to provide "a more appropriate distribution of public funds, and to create greater interest amongst Greenlanders in their own situation."[51] Under this system, Greenlandic men in each settlement elected a *forstander*, or local chief, who (with the council) met annually with colonial officials to discuss policies and to disburse funds for the poor. The councils were significant because they laid the groundwork for the development of Greenlandic-based institutions of governance, including Home Rule.

As the nineteenth century came to a close in West Greenland, local Greenlanders were well enmeshed in a transformed colonial society. They increasingly recognized a common national identity as *Kalaallit* and had established regular contacts with East Greenland and North Greenland. Their livelihood as hunters, however, was under increasing stress as world prices for seal and whale oil began to decline and as significant climatic changes began to occur.

Transformation of West Greenland's economy, 1900 to World War II

During the early twentieth century, dramatic environmental changes transformed the economic basis for much of West Greenlandic society from hunting to commercial fishing. A dramatic rise in ocean temperature about 1916 (see figure 3) brought a massive influx of cod from the North Atlantic. At the same time, seal populations shifted northward, reducing their availability. During this period, Greenland's population was increasing: From 1880 to 1920, it grew nearly 50 percent, from just below 10,000 to nearly 15,000.[52] Declining food production and loss of income from sealskins forced Greenlanders to look for new livelihoods. Although they had long been involved in fishing, this activity was primarily for local consumption and not for export.

The influx of cod led to the first of four major development phases in Greenland's commercial fisheries.[53] The first phase was preindustrial, which extended from the 1920s to about 1950. During this era, fishermen used small boats in summer to catch cod and preserve them with salt. A second phase developed in the 1950s and 1960s when fishermen delivered to small, land-based processors. During this period, fishermen also began to catch shrimp and salmon for export. The third phase extended from the 1970s to the mid-1980s, when large, ocean-going trawlers took shrimp in increasing amounts and cod catches began to decline. During this phase, authorities made major investments in fishing vessels and in on-shore processing facilities. In the 1990s, cod catches dropped to very low levels, and fishermen became even more dependent upon shrimp. Fishermen began searching actively for other species, and exports of Greenlandic halibut and crab began to expand.

The transition from a hunting to a fishing economy in West Greenland was not always easy. Those who grew up as seal hunters were often reluctant to take part in commercial fisheries. As Greenlander John Møller wrote:

At first, one fished from a kayak, and then from a rowboat, and when it became possible to buy a motorboat on credit, the way was paved for the fishery to become a real occupation. In the beginning, the dedicated hunters refused to give up their long-standing occupation and defended it by tooth and nail. . . . The reaction can best be compared with the opposition in 19th century Europe to achieving greater efficiency through technological development.[54]

When Denmark's policy of protecting hunting livelihoods changed to the promotion of commercial fishing, the technologies available to Greenlanders in the 1920s and 1930s also changed. The number of European-style fishing boats increased dramatically, and a corresponding decline occurred in the use of kayaks and *umiat*.

In addition to fisheries, colonial authorities encouraged other forms of economic development, including sheep ranching in southern Greenland and an expansion of community-based whaling. Until the 1920s, hunters in Paamiut and Nuuk continued to catch mostly humpback whales. After World War I, Danish authorities expanded whaling efforts on behalf of Greenlanders using a British-built vessel called the *Sonja*.[55] The vessel's Danish crew caught fin, blue, and other whales and delivered them to local communities. The meat and *mattak* were consumed locally, while the blubber was shipped to Denmark for processing (see chapter 4). Mining for cryolite, a mineral used in producing aluminum, provided a major source of income for Greenland. The substance had been mined at Ivittuut in West Greenland since 1859, but it took on new importance when aluminum became used for aircraft construction.

Economic diversification was accompanied by significant changes in colonial governance. In 1905, the Danish *Folketing* (parliament) formally disbanded the state-sponsored mission. In 1908, it adopted a new administrative law that replaced the two inspectorates for northern and southern Greenland with two provincial councils (Danish, *landsråd*), which had advisory powers. The 1908 law also created local "municipalities" (Danish, *kommuner*). These changes enabled Greenlanders to exert greater control over local affairs and established the basis for the system of municipalities that continues today.

World War I only minimally impacted Greenland, although it signaled the growing significance of world events on Greenlandic-Danish relations. In 1916, Denmark sold its interests in the Virgin Islands to the United States, leaving Greenland, Iceland, and the Faroe Islands as its only remaining colonies. Following the war, Denmark consolidated its trade monopoly over all of Greenland. In the process, Danish authorities alienated many Greenlanders in 1924 by signing a Treaty of East Greenland with Norway. The treaty granted Norwegians hunting rights in the region but left undecided key questions about who had sovereignty over the region. Greenlanders were shocked that the Danish government acceded unilaterally to the Norwegian demands, but they were doubly irritated over Denmark's failure to consult with the provincial councils. This action led many Greenlanders to reassess attitudes toward Denmark. According to Petersen,

Seen from a Greenlandic perspective, [Danish-Greenlandic] cooperation was crucial, but the dismay [at the Danish decision] was enormous. It became clear that it was a striking lack of knowledge of Greenlandic conditions that led the Danish government to negotiate as it did. In Greenlanders' eyes, Denmark had made a serious error.[56]

Greenlanders' insistence that they be active participants in decisionmaking was reflected in a new administrative law adopted in 1925. The new law retained the administrative division between northern and southern Greenland but modified the governance system by establishing district councils. These councils were comprised of municipal representatives and colonial officials, who were to review and comment on all proposed changes in Danish law affecting Greenland. Despite these changes, KGH continued to dominate Greenland's economy and to thwart any significant buildup of private capital.

The G-50 and G-60 eras: Social experiments in Greenland

Greenland was almost cut off completely from Denmark during World War II. The Nazi invasion of Denmark in 1940 served as a potent reminder of Greenland's vulnerability and strategic significance. Because Greenland's ties to Europe were disrupted, the Danish ambassador in the United States concluded a treaty allowing Allied forces to build military bases there. Greenland's cryolite resource became vital to wartime production of aluminum, and the country's strategic location in the North Atlantic proved crucial to protecting Allied shipping to Europe.

Danish authorities in Greenland sought to retain a semblance of normality in the country during the war, but the influx of American troops and goods made a significant impact on Greenlanders' self-awareness; they were no longer exclusively dependent on the Danes.

Greenlanders began to realize that greater self-government was not simply a dream but was within reach.

Following the war, Danish authorities and Greenlanders embarked on an effort to "normalize" relations between Denmark and her colony. This process had far-reaching impacts on Greenlandic life in the years ahead. Local leaders challenged the dominance of KGH and pressed for the creation of a single governing council for the entire country. In 1946, they presented these ideas to a joint Greenlandic-Danish committee charged with developing a five-year plan for the country's future. The committee, however, rejected any fundamental changes to KGH and dismissed the idea of a single council, fearing that it would advance "separatist" ideas.[57]

The committee's actions were poorly received in Denmark, where public sentiment had become staunchly anticolonial. Political pressure forced Danish politicians to push for more dramatic changes. As a result, the government formed the so-called "Big Commission" in 1948, consisting of four Greenlanders and four Danes. The commission's recommendations, published in 1950, addressed issues ranging from education to economic development to administrative reform. They formed the basis for a ten-year state modernization plan, popularly known as G-50. This plan envisioned a continuing transformation of Greenland's economy through massive investment in new docks, fish-processing plants, and housing in the open water districts of southern Greenland. Its far-reaching impact on Greenlandic society was not lost on commission members. The Danish chair noted:

One should not forget that such a conscious and intended transformation of a society . . . means intervening in a most decisive way in the existence of each and every human being. From certain points of view, one might say that it is a social experiment on a large scale, but that urgently requires the utmost attention and care in implementation. And it must above all be avoided that the endeavor to develop the economic, cultural, and social conditions acquires a convulsive character.[58]

Two of the most significant results of the G-50 plan were the abolition of the KGH trade monopoly and the creation of a single provincial council for all of Greenland, based on Nuuk. The shift to a single council reflected a growing sense of national identity among Greenlanders, who recognized their common interests in dealing with Denmark. This solidarity was tested by the arrival of large numbers of Danish construction workers, whose job it was to build new housing and infrastructure. The Danish population increased rapidly, with most workers taking well-paying jobs in construction and government. Greenlanders increasingly felt like strangers in their own land.

Greenland's status as a Danish colony ended in June of 1953, when Danish voters amended the national constitution. In theory, Greenland became an integrated part of Denmark, and Greenlanders thereafter were to become "northern Danes," entitled to all the rights and privileges of full citizenship. Dahl, however, argues that this transition brought on colonialism in a new guise, even though many Greenlandic leaders supported these initiatives.[59]

Frustration with the results of G-50 led to creation of another state-controlled commission, the Greenland Committee of 1960 (referred to as G-60). The G-60 commission's goal was to normalize relations between Greenlandic and Danish institutions; in large measure, to assimilate Greenland fully into the Danish realm. Greenlandic leaders pressed to give the provincial council more authority, to reduce the power of Danish bureaucrats, and to encourage greater private investment. As a result, investment in Greenland increased over tenfold between 1960 and 1980.[60]

Although investment increased rapidly under G-60, the plan's contradictions quickly became apparent. Its ambitious agenda of social engineering created a vicious cycle. As more Danish workers poured into Greenland to build new infrastructure, they demanded better housing and services than Greenlanders themselves had. Resentment grew as disparities in income between Danes and Greenlanders became more apparent. In 1967, Danes in Greenland comprised only 15 percent of the population yet earned 50 percent of all private income.[61]

A key element in the G-60 plan was the resettlement of Greenlanders from smaller settlements to larger towns, where they would work in fish processing. Although most Greenlanders were not forced to move, many were compelled to do so after authorities shut down services such as post offices, schools, and stores. This resettlement scheme had profound impact on West Greenland: The proportion of Greenlanders living in larger towns grew from about 58 percent in 1960 to over 80 percent in 1990. In the process, many hunters lost the basis for procuring country foods for their own households.

Both the G-50 and G-60 plans were designed around improving Greenland's cod fishery. Ironically, just as these plans were coming to fruition, cod stocks off West Greenland began to decline. As a result, economic initiatives shifted from smaller coastal fisheries to high-seas trawling in Davis Strait and beyond. This move required even greater capitalization and use of advanced technologies, thus increasing dependence on imported capital and expertise. KGH, still a powerhouse in Greenland's economy, purchased seven large trawlers in the late 1960s. By the mid-1970s, KGH's vessels produced over one-third of all fish products in Greenland, including over one-half of cod resources. As Dahl observes, "for the first time in the 250-year-long colonial history, KGH now dominated the entire process, from the time products were caught to when they were sold in Denmark and other lands."[62] The rapidity of these changes led to new tensions, and Greenlanders increasing called for development that would reflect what became known as the "special Greenlandic conditions." This movement set the stage for dramatic changes in the 1970s, leading to the creation of Home Rule.

The struggle for Home Rule, 1970–1979

The uneven development of the G-50 and G-60 eras was underscored in the early 1970s by growing political tensions. The decision of Danish authorities in 1972 to shut down Qullissat, a community on Disko Island built around an unprofitable coal mine, sparked a outpouring of Greenlandic nationalism. Greenlanders viewed this action as an indictment of G-60 resettlement programs and postcolonial policies.

The Danish government also generated controversy by deciding in 1975 to offer offshore oil and gas concessions in Davis Strait to multinational corporations. Greenlandic politicians objected strenuously, arguing that Greenlanders had aboriginal rights to these resources. As it turned out, exploratory drilling failed to produce commercial quantities of oil and gas, but the controversy highlighted a rapidly developing conflict over ownership of nonrenewable resources.

The Home Rule process was advanced in the 1970s by emergence of a young and radicalized Greenlandic elite. Greenlandic students studying at Danish universities were captivated by the ideological fervor of the times. Young Greenlanders like Jonathan Motzfeldt and Moses Olsen turned their efforts toward creating "a more Greenlandic Greenland." Members of *Peqatigiit Kalaallit* ("The Young Greenlanders' Association") in Copenhagen combined nationalist and anticapitalist ideologies to press for greater Greenlandic self-determination. Lars Emil Johansen, now Greenland's prime minister, said in 1971 that "Greenlanders and Danes in Greenland must acknowledge the fact that we are two different peoples, and that we must not aim for integration of the two. . . . Only on this basis can we achieve cooperation that builds on mutual respect."[63] The elite's influence was magnified in 1971 when Moses Olsen was elected to Greenland's seat in the Danish *Folketing*. As events unfolded, he ended up holding a swing-vote in an otherwise deadlocked parliament. This situation provided Greenlanders with extraordinary leverage in pressing for Home Rule.

Differences between Danish and Greenlandic attitudes became even clearer during a 1972 referendum on Denmark's membership in the European Community (EC). Danes supported joining, but Greenlanders voted over two-to-one against. The major issue was concern about fishing rights for EC nations in Greenlandic waters. This mobilization contributed to the formation of Greenlandic political parties, which continue to play a significant role in political life today. The *Siumut* party (Greenlandic for "forward") began in the early 1970s and was incorporated as a moderate socialist party in 1977. The *Atassut* (or "connections") party was also formed in 1977 but had a more conservative-liberal platform supportive of strong ties with Denmark. The other major party, *Inuit Ataqatigiit* ("human brotherhood"), was to the left of *Siumut* and favored complete independence for Greenland.

Political pressure finally led the minister for Greenland to appoint a committee of Greenlanders to make recommendations about enhancing self-determination. Adapting a model of Home Rule used earlier in the Faroe Islands, the committee called for creation of a Commission on Home Rule in Greenland. This commission began work in 1975 and consisted of seven Danish and seven Greenlandic members. It completed its work in 1978, recommending a system of Home Rule that retained the unity of the Danish realm. Greenlanders' aboriginal rights were hotly debated by the commission, with *Siumut* representatives arguing that Greenlanders had full and complete ownership of nonrenewable resources under international law. Negotiations nearly broke down until Denmark's prime minister stated bluntly that Greenland could insist on full ownership, but doing so would mean leaving the Danish realm. Greenlanders, recognizing their continuing dependency on Denmark, were forced to back down. The parties compromised on language affirming Greenlanders' "fundamental rights" to the natural resources of Greenland: a nebulous and as-yet-untested legal concept. Nevertheless, as Kleivan points out,

Home Rule is the application to the indigenous people in Greenland of the general principle of international law called self-determination . . . [but] the negotiations in the Commission on Home Rule, and the Home Rule Act itself are a strong reminder that "self-determination" is not a clearly defined principle. Transfer of the right to self-determination through a home rule scheme can be designed with varying degrees of respect for the indigenous population. [However,] the arrangement for Greenland leaves ample room for the state authorities to safeguard state interests and control.[64] [references omitted]

Namminersornerullutik Oqartussat: *The Home Rule state*

In contemporary Greenlandic discourse, "Home Rule" has many meanings.[65] When Greenlanders speak of Home Rule, they are likely referring to the Home Rule government itself—to the power, apparatus, and functions of political institutions. But the concept also has broader meanings: language and cultural revitalization, nation-building, and growing international ties to Inuit in Canada, Alaska, and Chukotka. Although the institutions of Home Rule are largely in Nuuk, the Home Rule state includes an extensive "subgovernment," made up of Greenland's eighteen municipalities as well as major unions, business groups, and social organizations. Each of these entities plays a role in shaping policies affecting Greenland's future, including whaling.

The Danish *Folketing* formally approved Greenlandic Home Rule on November 29, 1978. Home Rule became a reality on May 1, 1979, after Greenlandic voters approved it overwhelmingly in a referendum. This action established a popularly elected parliament, the *Landsting*, and the Home Rule government itself, the *Landsstyre*. In the West Greenlandic language, these entities are collectively referred to as *Namminersornerullutik Oqartussat*, or "Home Rule." The *Landsting* consists of thirty-one members, each elected for a four-year period (although elections can be called at any time). The *Landsting* usually meets twice a year, once in the spring and then again in fall. The majority in the *Landsting* chooses the government itself (figure 2). This government consists of a premier and six other members, who may or may not be elected officials. Greenland's current premier is Lars Emil Johansen, the leader of *Siumut*. The current *Landsstyre* is made up of a coalition of *Siumut* and *Atassut*. Under Home Rule, Greenland continues to be represented in the Danish parliament by two elected representatives.

The Home Rule Act divides authority for the Home Rule and Danish states into three areas: (1) those fully within the Home Rule's com-

PREMIER Lars Emil Johannsen

Responsibilities: Leader of Home Rule government, relations with Danish authorities, elections, international relations, supervision of municipal governments, nonrenewable resources, legal office, planning for villages, administrative office

* * *

MINISTRY OF FINANCE AND HOUSING
Daniel Skifte

Responsibilities: Budgets and accounts, financial administration systems, investment planning, personnel, town and regional planning, taxation, statistical office, computer systems, planning for housing, housing administration and support

* * *

MINISTRY OF FISHERIES, HUNTING, AND AGRICULTURE
Pâviâraq Heilmann

Responsibilities: Management of renewable resources & fisheries, hunting, and agriculture

* * *

MINISTRY OF HEALTH, ENVIRONMENT, AND RESEARCH
Marianne Jensen

Responsibilities: Health care system, alcohol policies, workplace issues, environmental conservation, Greenland Institute of Natural Resources, marine affairs, research policies, Nordic Council relations

* * *

MINISTRY OF CULTURE, EDUCATION, AND RELIGIOUS AFFAIRS
Konrad Steenholdt

Responsibilities: Public schools, higher education, vocational training, folk high schools, libraries, museums & archives, cultural issues, media relations, church affairs

* * *

MINISTRY OF SOCIAL AFFAIRS AND LABOR
Benedikte Thorsteinsson

Responsibilities: Social affairs, social security, social services planning, residential institutions, employment services

* * *

MINISTRY OF INDUSTRY, TRANSPORT, AND PUBLIC WORKS
Peter Grønvold Samuelson

Responsibilities: Industrial development & policy, trade, service sector, tourism, international trade, business policies, consumer affairs, transportation and technical services, telecommunications, energy, building and construction, airports authority, shipyards, information technology

FIGURE 2. Greenland's *Landsstyre* (executive branch), 1995[66]

petence; (2) those fully under the Danish state; and (3) those shared by both governments. Since 1979, the Home Rule government has gradually assumed control over nearly all internal matters, ranging from fisheries management to education and health services to development of economic infrastructure. The most recent takeover was the health care system in 1992, which was both the most complex and most expensive undertaking.

Under Home Rule, the Danish state retains authority over foreign policy, defense, fiscal and monetary policy, and the justice system. In addition, the Home Rule relationship implicitly retains a number of provisions common to all parties within the Danish realm, including acceptance of the Danish constitution, the authority of the Danish *Folketing*, and Danish law relating to human and individual rights.

Controversy in the 1970s and 80s over Greenland's continuing relationship with the European Community—now the European Union (EU)—tested provisions of the Home Rule Act regarding foreign policy. Greenlanders' resistance to membership was clear from the 1972 referendum; they clearly feared that Brussels would dictate fisheries policy. The issue of the country's continuing membership came to a head in a 1982 election, which proved to be a major test of Home Rule. Voters decided overwhelmingly to withdraw from the EC. Despite this decision, Greenland has been able to retain favorable trading status, which brings lucrative multiyear fisheries agreements and free access to European markets. Denmark's continuing control over foreign policy also means that the American military base at Pituffik (Thule), built in the early 50s, continues to operate despite Home Rule concerns. Denmark remains a part of NATO and views its agreement with the United States as part of its continuing obligation to a common defense strategy.

The principal area where Greenland and Denmark share authority is in the management of nonrenewable resources. Although the Home Rule Act recognizes Greenlanders' "fundamental rights" to these resources, actual decisionmaking about mineral exploration and development is in the hands of a joint commission with equal Danish and Greenlandic membership.[67] The *Landsting* retains veto authority over all commission decisions. Under new procedures adopted in 1988, any income from hydrocarbon or mineral development up to 500 million DKK (about $90.9 million) is shared equally between the Danish and Home Rule governments. If income exceeds 500 million DKK, the two governments will renegotiate how the funds will be allocated. Importantly, unlike an earlier agreement, income received by the Home Rule government does not affect the amount of Greenland's annual subsidy from Denmark.

A second major area where Greenland and Denmark share control over resources is with regard to management of fin and minke whaling, subject to the jurisdiction of the IWC. I return to this topic in depth in later chapters.

Municipal governments also play a major role in Greenlandic life. Greenland's eighteen municipalities (Danish, *kommuner*) administer a wide range of local services and have the power of taxation. They are governed by an elected municipal council, which selects a mayor. Among other responsibilities, municipal councils can enact regulations regarding hunting and fishing beyond those enacted by the *Landsting*. The smallest settlements also have their own local councils (*bygderåd*). Municipal governments in Greenland are represented at the national level by KANUKOKA, *Kalaallit Nunaani Kommuneqarfiit Katuffiat* (the "Association of Greenlandic Municipalities"), formed in 1972.

Greenland's political system also includes organizations like KNAPK, or *Kalaallit Nunaani Aalisartut Piniartullu Katuffiat* (the "Organization of Hunters and Fishermen in Greenland"). KNAPK is an umbrella organization representing seventy-one local hunters' and fishers' organizations throughout Greenland. Based in Nuuk, it has a board of twelve members elected from seven districts. KNAPK negotiates price agreements for hunting and fishing products with Royal Greenland, the Home Rule–owned fisheries company, and is a significant player in debates about whaling policy.

At a Crossroads: Greenland's economy in the 1990s

Greenland today is at an economic crossroads. Like many countries in the developing world, Greenland is seeking to diversify its economy and to expand market principles in economic life. Until recently, Home Rule development policies focused on "Greenlandization"; shifting control over major forces of production (particularly fisheries) and government services from Danish to Greenlandic hands. That goal was largely achieved by the early 1990s. At the same time, Greenland's leadership did away with the last vestige of economic discrimination separating Danes and Greenlanders—the so-called "birthplace criteria," where the former received greater compensation for nearly identical work.

The challenge for Home Rule in the 1990s is to strengthen Greenland's economy. Today that economy is highly dependent upon shrimp exports and Danish subsidies. In 1995, Greenland's gross domestic product amounted to about 6.37 billion DKK, or about $1.16 billion. Fish products provide virtually all of Greenland's export income, however, and nearly 80 percent of that income is from one product: cold water shrimp. In 1993, Greenland caught 80 thousand metric tons of shrimp, making it the world's largest exporter. Biologists believe that shrimp stocks—indeed, many fisheries stocks in Greenlandic waters—are being utilized at their maximum sustainable level.[68] Greenland's dependency on fisheries has grown as mining and other income sources have declined (figure 3). At the same time, income from hunting (primarily sales of sealskins) has declined steadily in recent years to the point where it now comprises less than 1 percent of GNP.[70]

Danish subsidies to Greenland are substantial, amounting to nearly 50 percent of GNP, or about 3.3 billion DKK annually (about $545 million) (figure 4). In most cases, these subsidies are tied to specific services (e.g., the health care system) handed over by the Danish government to the Home Rule government. Today, the level of subsidies continues to be a major issue in Danish-Greenlandic relations, and a few Danish political leaders advocate significant reductions.

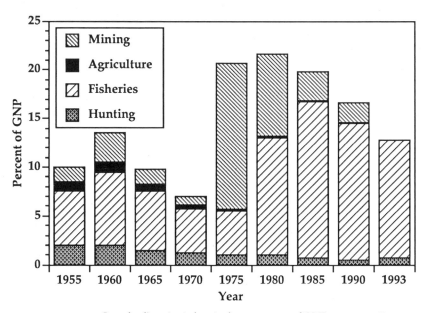

FIGURE 3. Greenland's major industries by percentage of GNP, 1955–1993[69]

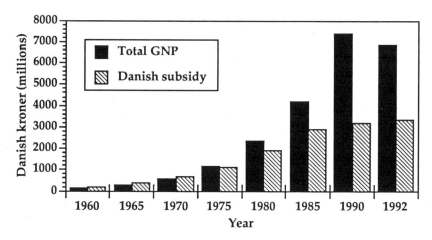

FIGURE 4. Danish subsidy as proportion of Greenland's GNP, 1960–1992[72]

Critics argue that these subsidies distort Greenland's economy and undermine private initiative. Without subsidies, they argue, Greenland would have a GNP about that of Portugal, one of the EU's poorest countries.[71] With subsidies, its per capita GNP approaches that of Denmark itself. Subsidies also support an unusually large public sector with substantial administrative costs. Uncertainty about Denmark's willingness to continue these subsidies in the years ahead adds to Greenland's economic vulnerability. This concern became all too clear when, in 1987 and 1988, the Home Rule government was forced to borrow nearly $150 million from Japanese and other lenders to avert a cash flow crisis.

Recent debates about Greenland's economic future were sparked by a book produced in 1994 by Danish economist Martin Paldam that compares Greenland's economy with that of developing countries.[73] Paldam called for greater privatization and significant reductions in government expenditures. He also cast doubt on the future of small-scale hunting and fishing in Greenlandic life and questioned the costs of subsidizing smaller settlements. His book was greeted with criticism in Greenland in part because it overlooked what many believe to be an essential part of Greenlandic life—the contribution of hunting and settlement life to contemporary livelihoods. For example, Paldam argues that the contemporary hunting sector is economically "irrational":

[T]here is almost no net return from the hunting industry. . . . [W]e have here a segment of the population that can have an acceptable standard of living *only* by subsidizing it heavily. In this sense, we can say that we have here a sort of regulated *museum* industry; . . . it represents a connection to the past that one will not sever because of its connection to a collective identity.[74]

Data about household economies presented in chapter 2 of this book raise doubts about Paldam's argument. By analyzing Greenland's economy in conventional economic terms, he overlooks the continuing importance of household-based production of country foods in many communities. He moreover ignores the underlying reason for declines in markets for sealskins; that is, anti-sealing campaigns in Europe and North America that undermine what could be a significant source of income for Greenlandic hunters.

This on-going debate is part of a broader Home Rule initiative for economic restructuring now underway in Greenland. The initiative has five major goals: (1) greater economic diversification; (2) privatization and reduction of the public sector; (3) development of a more cost-effective infrastructure; (4) improved management of renewable resources (especially fisheries); and (5) balancing development between larger towns and smaller settlements.[75] Greenlandic leaders are promoting mineral exploration, expanding tourism, and developing new fisheries and land-based industries. Greenland currently has no operating mines, but intensive exploration is underway for both hydrocarbon resources and minerals.[76] Some of the most promising efforts involve exploratory drilling for oil and gas on the Nuussuaq Peninsula near Disko Bay and assessment of a large zinc deposit in Citron Fjord, North Greenland. There are also sizable gold deposits at Kangerlussuaq, East Greenland, and near Nanortalik. Home Rule officials are working with the Canadian multinational Platinova, Inc., to determine the costs and benefits of a zinc-processing plant near Nuuk. Premier Lars Emil Johansen believes this project to be one of the most hopeful for Greenland's long-term economic future:

I fervently hope that our efforts in developing mineral resources will soon bear fruit in the form of new mining enterprises. . . . A [zinc processing] facility of this magnitude can give [Greenland] the economic shot in the arm that the Landsstyre is working so hard to achieve.[77]

Tourism is an increasingly important element in Greenland's economic future. Greenland currently receives about five to seven thousand tourists per year. Its tourism potential is hampered by high costs, a short season, difficult and costly flight connections to other countries, and a poorly developed infrastructure. The Home Rule government has an ambitious plan to increase tourist visits to 35,000 per year by the year 2005. Proponents believe that tourism could create as many as two thousand new jobs and bring in as much as 500 million DKK (about $90 million) annually.

Economic restructuring also involves cutting government and privatizing state-owned enterprises. According to Premier Johansen:

the [Greenlandic Parliament] is prioritizing growth over welfare when decisions are made about new investment. And in recognition of this, we have taken steps to restructure Home Rule enterprises into competitive businesses. This is absolutely necessary in order for them to survive in the international market.[78]

Privatization began with a major restructuring of Home Rule–owned Royal Greenland in 1989 and 1990. This action proved to be an impressive success. In 1991, Home Rule politicians also broke up state-owned *Kalaallit Niuerfiat* (KNI) into smaller, for-profit companies, with responsibilities for transportation, wholesaling, and retailing. This move has been much less successful, and the Home Rule government has had to cover significant losses during the transition period.

Home Rule leaders are also working to make Greenland's infrastructure more cost effective. The most significant effort now underway involves construction of seven new runways in larger Greenlandic towns. Once completed, the runways will enable Greenland Air to operate fixed-wing aircraft rather than the aging helicopters now in use. Nuuk's existing runway might also be expanded to accommodate jets serving Europe and North America. Another major investment is a large hydroelectric facility near Nuuk. The project, which came on line in 1993, cost just over $1 billion and was financed by the Home Rule government and Norwegian lenders. It is already reducing Greenland's dependence on imported oil, and it opens up new opportunities for industrial development near Nuuk.

Economic restructuring also involves improving management of fisheries and other renewable resources. In the late 1980s and early 1990s, biologists pressed for reductions in the overcapitalized shrimp fishery. They believed that shrimp quotas should be reduced from current levels of over 80 thousand tons per year to about 50 thousand tons. To accomplish this reduction, the Home Rule implemented a system of transferable vessel quotas in the offshore shrimp fishery in 1991. Plans call for continuing reductions of shrimp catches, amounting to 5 percent annually until 1998, to ensure sustainable catches. Improved fisheries management required renegotiation of fisheries agreements with the EU, Norway, Russia, and other countries. Greenland recently signed a new agreement with the EU for a six-year period beginning in 1995. It allows EU vessels to catch cod, redfish, and other species and generates some 283 million DKK (over $51 million) annually for Home Rule coffers.

The final element in economic restructuring focuses on balancing development between larger towns and smaller settlements. This political issue is a delicate one for Greenlandic politicians. Public sensitivity to the needs of settlements remains high, and many outside of Nuuk express concern about the capital's dominance in economic life. Home Rule officials are responding by moving some state-controlled businesses to regional centers to distribute investments and jobs more widely. Some critics, however, continue to be concerned about a highly centralized economy in Greenland—even under Home Rule—and point to similarities with earlier G-50 and G-60 eras:

G-60 was a perfect example of what can happen when someone sits down at a desk and figures out what is best for the Greenlandic people, and then forces their plan through. This time, the desk is in Nuuk, and "hurrah" for that! But it is just for that reason that it is inconceivable that politicians now show the same open contempt for people's right to decide for themselves when it comes to major decisions of fundamental importance to Greenland's future.[79]

Another Greenlander, writing about the future of smaller settlements, commented:

I can well understand why those who live in settlements and outlying areas are disappointed with development, and that they are unsure about the future. The Home Rule Government has not lived up to the promises it gave when it was put into place. . . . We are spectators to the development of our own land. Twenty people here in Greenland plan and decide what will happen to 50,000 people.[80]

Greenlandic political leaders don't shy away from this criticism. Jonathan Motzfeldt, formerly Greenland's premier and currently head of KNI, emphasizes that

we live in a world which is more in flux than ever before, and where the labor force moves more and more across borders. . . . That is healthy, and we Greenlanders also have our place in the world. Naturally we shouldn't uncritically import workers, but neither should we simply say "Greenland for Greenlanders." That would isolate us from the rest of the world and create a reservation only for Native people. . . . Not only is our economic condition dependent on the rest of the world, but politically the circumstances and changes have a direct influence on us.[81]

These contradictions in Greenland's economy and their impact on whaling issues become even clearer when we examine changing production relations at the community and household levels. Chapter 2 focuses on these relations in Qeqertarsuaq Municipality in West Greenland, revealing the characteristics of the kin-ordered mode of production there. This analysis provides an essential backdrop for understanding the economic, social, and cultural significance of whaling in Greenlandic life and highlights relationships between community-based whaling and distant political and economic forces.

Changing Production Relations in Greenland: The Case of Qeqertarsuaq Municipality

This chapter is a case study of changing production relations in Qeqertarsuaq Municipality in northern West Greenland (map 2). Before we examine whaling practices, we need first to understand the mode of production in contemporary Greenlandic communities. As we shall see, the economies of these communities are no longer exclusively subsistence oriented, nor are they preindustrial in character; rather, they are based upon a mixed subsistence-cash economy where small-scale production for market exchange (so-called simple commodity production),[1] wage employment, and transfer payments complement a household's own production of wild foods. Researchers working in northern communities in Alaska and Canada have identified at least six characteristics of this mixed subsistence-cash economy:[2]

- community-wide, seasonal round of production activities
- high household production of wild resources
- primarily kinship-based social organization for production
- extensive noncommercial distribution and exchange networks
- traditional systems of land use and occupancy
- cash used to support household hunting and fishing

By focusing on these characteristics in Qeqertarsuaq, we begin to understand the significance of kinship and alliances in organizing production relations, the role of cash in household reproduction, and the continuing importance of cultural beliefs and practices. The data presented here also provide a backdrop for understanding whaling's significance in Greenland's mixed economies, which I discuss in depth in chapter 3.

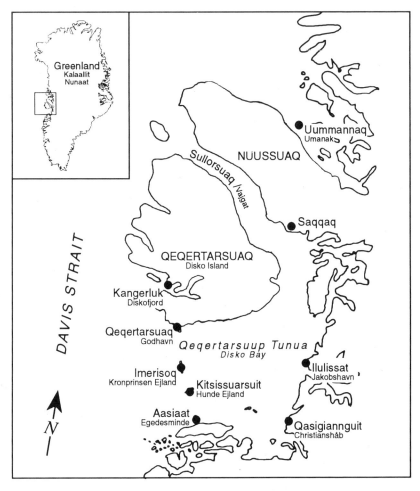

MAP 2. Qeqertarsuaq Municipality and the Disko Bay region

Qeqertarsuaq Municipality and the Disko Bay region

Qeqertarsuaq Municipality provides a particularly useful case study for understanding whaling because of its rich history as a whaling center and its dynamic mixed economy. Its 1,191 residents live in two communities: the town of Qeqertarsuaq itself (population 1,125) and the smaller settlement of Kangerluk (population 66).[3] Nearly all inhabitants (90 percent) are *Kalaallit*, or indigenous Greenlanders, and virtually all speak *Kalaallisut* as their first language. Danes make up the remainder of the population. Today, Qeqertarsuaq's economy is based on commercial fishing, public sector employment, tourism, and household-based fishing and hunting. Households with relatively high incomes—the so-called "millionaires" of the shrimp industry—exist alongside those with relatively low incomes and high subsistence production. In this sense, Qeqertarsuaq combines characteristics of predominantly fishing communities to the south (e.g., Qaqortoq and Paamiut) and those to the north (Uummannaq and Upernavik), where hunting is paramount. Most importantly for our purposes, whaling in Qeqertarsuaq continues to be an important part of people's history, economy, and identity.

The political boundaries of the municipality encompass all of Qeqertarsuaq Island on the north side of Disko Bay and a small island group in the bay itself.[4] The island is Greenland's largest (8,578 square kilometers), and its higher elevations—reaching over two thousand meters—are covered with glaciers and permanent icefields. At lower elevations, it has an unusually diverse array of vegetation, contributing to its reputation for scenic beauty. Disko Bay is over eight hundred meters deep and has an abundance of shrimp, fish, and whales. Nearby Jacobshavn Glacier regularly produces enormous icebergs visible from Qeqertarsuaq throughout the year. Sea ice covers most of the bay from December until April. The local climate is maritime. The average temperature is $-2.5°C$, but temperatures can reach $18°C$ in summer or drop as low as $-30°C$ in winter. In midwinter, the sun drops below the horizon for six weeks; in summer it is above the horizon twenty-four hours a day.

Marine resources dominate economic life in Qeqertarsuaq. Shrimp, Greenland halibut, and (increasingly) crab are the most valuable commercial species, but fishermen also take salmon, Atlantic cod, lumpfish, Greenland shark, Atlantic halibut, and Arctic char. Hunters take a wide variety of marine mammals, including the ringed seal, harp seal, hooded seal, beluga, narwhal, harbor porpoise, minke, and fin whales (figure 5). Polar bears are also occasionally hunted. Land mammals are far less common. Hunters travel long distances by boat to take caribou and muskox, and catches are limited. Hunters also catch eider ducks, murres, kittiwakes, and ptarmigan.

The town of Qeqertarsuaq has a shrimp processing plant and a small facility for handling hunting and fishing products, both owned by Royal Greenland A/S. It also has a large dock and warehouse, one ma-

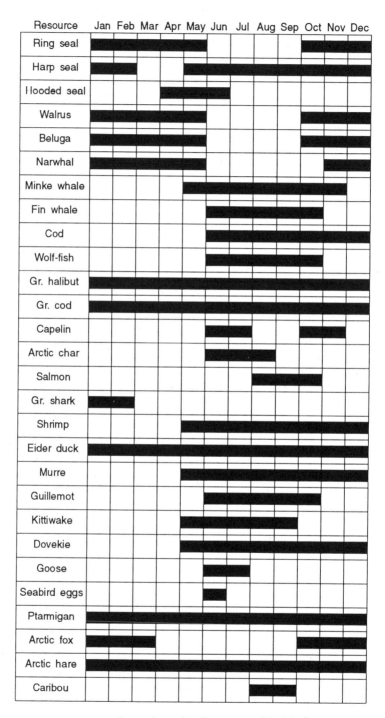

FIGURE 5. Seasonal round in Qeqertarsuaq Municipality

PLATE 1. View of Qeqertarsuaq in summer

jor store (selling food, furniture, building supplies, and clothing), a bakery, several smaller convenience shops, a small hospital, banking and ticket offices, a post office, a church and museum, and a heliport. The University of Copenhagen operates the Arctic Station in Qeqertarsuaq, a year-round research facility employing several people. The municipality and the Home Rule government own or have a controlling interest in most of the major infrastructure in the town.

Qeqertarsuaq has several hundred dwellings owned by the Home Rule–controlled housing company, by the municipality, and by private families. These buildings include single-family homes, duplexes, and apartments. Nearly all residences have electricity and running water. A public utility produces electricity with diesel generators. Water is piped in from a nearby river, although in winter some households have it delivered by truck. Sewage is gathered in "honey buckets" and collected by the municipality on a regular basis. Nearly all homes in Qeqertarsuaq have telephone service, and all receive limited radio and television programming.

Kangerluk is a much smaller community, with only fifteen homes, a chapel, one small store, and a post office. Royal Greenland owns a small processing facility there, used for purchasing dried fish and meat. The town has a diesel generator that provides electric power, and water is delivered to local homes. As in Qeqertarsuaq, sewage is collected in honey buckets. Kangerluk only has a single community telephone located in the store. Residents receive regular radio programming but have tape-delayed television.

Most goods sent to Qeqertarsuaq and Kangerluk travel by ship, which arrive regularly during the ice-free period. A smaller KNI vessel carries passengers and freight from Qeqertarsuaq to Kangerluk from June to November. Coastal passenger ferries visit Qeqertarsuaq town about every ten days during the open-water season. Greenland Air provides year-round helicopter service to Qeqertarsuaq, but Kangerluk has no regular air service.

The municipality has a full-time mayor and an elected council. The latter is responsible for a wide array of local services, including schools, water and electrical systems, social services, fire protection, and cultural affairs. It also enacts local regulations governing hunting and fishing. Kangerluk residents also elect a local council. The municipality operates two schools, a day-care facility, and a senior citizens' home. Qeqertarsuaq's school has about 190 students and 20 teachers, while in Kangerluk there are about 14 students and 2 teachers.

A brief history of Qeqertarsuaq and Kangerluk

Qeqertarsuaq figures prominently in Greenlandic oral traditions, which recount how an *angakkoq* (shaman) of long ago towed the island with his kayak from southern Greenland to its present location to improve hunting opportunities.[5] Archaeological evidence from Disko Bay is limited, but a wide array of housepits, burial sites, and ruins suggest that the area was well populated in precolonial times.[6]

The first regular contact between Greenlanders and Europeans in Disko Bay was in the 1600s, when Dutch whalers and traders began making summer visits to the region. As European whaling began in earnest in the 1700s, Greenlanders increasingly sought out trading opportunities near Qeqertarsuaq, known then by Europeans as Liefde Bay. A visiting Danish missionary in 1738 found over two hundred people living in tents near the present-day community.[7]

When Greenland became a Danish colony in 1721, traders and missionaries became increasingly common in the region. In an attempt to restrict Dutch trading activities, Danish authorities established the colonial station of Godhavn in 1773. Sven Sandgreen, a Swedish-born merchant, arrived there to undertake whaling and trading on their behalf. He hired local Greenlanders for whaling ventures and paid them with portions of the catch. Initially, Danes carried out whaling using skinboats, but after 1774 they began using European sloops and equipment. Whaling activities increased during the 1770s, with Danish ships overwintering in Godhavn, Fortune Bay, and other nearby locations. In 1782, Danish officials established Godhavn as the northern seat of colonial administration.

Foreign whaling was interrupted from 1807 to 1814 while Europe was embroiled in war. When it resumed in 1816, catches proved disappointing, and efforts to revive the industry in the 1830s and 1840s brought little result. In 1851, Danish authorities finally shut down Godhavn's colonial whaling, citing its unprofitability. In the previous twenty-four years, only twenty-nine whales had been caught.[8]

Greenlanders in Qeqertarsuaq continued to catch Greenland right and other whales for local consumption after 1851, but on a sporadic basis. The colony's decline as a whaling station downgraded its economic and political significance; however, the town continued to be an important port-of-call for foreign whalers and polar explorers.

Climate changes and new fishing opportunities led to the creation of new fish salting plants in the communities in the 1930s.[9] These welcome additions followed a period during which sealing declined dramatically and income-producing opportunities were poor. During the 1920s and 1930s, Qeqertarsuaq also took advantage of whale catches by the vessel *Sonja*. During this period, the vessel delivered forty-three large whales to the community. New technologies also created change. The first motorized vessel arrived in Qeqertarsuaq in the late 1920s, and the use of kayaks and *umiat* began to decline in the 1930s and 40s.

World War II held up commercial production of salted fish in the area. During this period, households relied even more heavily on local hunting and fishing. After the war, the Greenlandic Fishing Company (*Det grønlandske Fiskerikompagni*) operated a freezer-ship for several years in Qeqertarsuaq, purchasing cod, wolf-fish, and Greenlandic halibut.

The G-50 plan resulted in the merger of the two provincial councils, and Qeqertarsuaq lost its status as Greenland's "northern capital"; however, the G-50 commission's plan led to major investments in Disko Bay fisheries, especially for shrimp. At the same time, officials shut down several outlying settlements. In 1965, Qeqertarsuaq's future was uncertain when a commission proposed that no new investments be made there, but events in the 1960s helped to revitalize the local economy. In 1962, fishermen succeeded in getting the processing ship *Sværdfisken* stationed in Qeqertarsuaq's harbor to purchase shrimp. In 1966, private investors built a small shrimp processing plant near the harbor. This construction enabled vessels to deliver locally, and it created jobs. In 1985, the plant was taken over by the Home Rule government, and it is now owned by Royal Greenland. In 1968, a local Greenlander also obtained government financing to build a small cold storage plant in Qeqertarsuaq for processing hunting products, especially whale meat and *mattak*.

In the 1950s and 60s, local fishing vessels began catching minke and fin whales. The first reported catch of a minke whale by a fishing vessel in Qeqertarsuaq was in 1958, when seven were reported taken.[10] In later years, annual minke catches varied considerably, but they reached as high as eighty-nine whales in one year. At about the same time, faster outboard motors made it possible for hunters in skiffs to participate in a collective hunt for minkes (see chapter 3). This activity gave households more direct access to whale products, because they did not have to purchase them from vessel owners.

During the 1970s and 80s, the price of sealskins plummeted as animal rights groups successfully campaigned against purchases of skins in Europe. Sealskin prices dropped from nearly 300 DKK in 1980 to 25 to 30 DKK in the 1990s. The Home Rule was forced to subsidize sealskin purchases to buffer hunters from depressed markets. During the 1980s, cod catches also declined dramatically, but shrimp catches increased as new stocks were discovered. Shrimp deliveries in Qeqertarsuaq increased from 883 metric tons in 1981 to nearly 2000 metric tons in 1990. The salmon fishery was also lucrative in the 1970s and 80s, but later declined with further quota restrictions.

In the 1990s, intense pressure on shrimp stocks and overcapitalization in the fishery caused Qeqertarsuaq residents to look for economic alternatives. A small crab fishery is now underway in Disko Bay, providing income opportunities both in summer and in winter. Tourism is also expanding. The region offers extraordinary scenic beauty, a rich history, and plentiful opportunities for hiking and other outdoor activities. The municipality recently opened a museum and a tourist office. Tourists arriving in June, July, and August can overnight at a small hut near the island's icecap and enjoy dog sledge and snowmachine trips under the midnight sun. A few tourists also come in the spring to take part in dog sledge expeditions.

Qeqertarsuaq's mixed economy: The wage sector

Households in Qeqertarsuaq earn cash in a variety of ways, including simple commodity production (primarily commercial fishing), wage employment, operating small businesses, and government transfer payments. Fish processing (again primarily for shrimp) and public employment are the two most significant sources of wages. In 1992, the average per capita wage income in the municipality was about $25,364 (139,499 DKK), somewhat higher than the average for Greenland of $23,723 (130,475 DKK). The largest wage employers are the municipality itself, the Home Rule government, and Home Rule–controlled enterprises, including Royal Greenland and KNI Pilersuiffik. In 1993, there were some ninety full-time public employees in Qeqertarsuaq and about seventy employees with KNI. Royal Greenland's shrimp processing plant employs ten people on a year-round basis but also hires over two hundred part-time employees for shrimp processing. In Kangerluk, Royal Greenland has only two employees. Most part-time employees are women. Interestingly, Royal Greenland often has difficulty hiring enough part-time employees during the peak summer season, because employees prefer to participate in hunting and fishing.

Unemployment figures for the municipality reflect the seasonal nature of wage employment (figure 6). Unemployment is highest in the first quarter of the year (January through March), when winter cold precludes fish processing and construction work, and is lowest during the busy summer season. Greenland's unemployment and social welfare programs are extensive. In Qeqertarsuaq in 1989, for example, these payments totaled about $1.7 million (9.33 million DKK).

Simple commodity production: Shrimp and other fisheries

In addition to wages, local households earn significant income from simple commodity (fisheries) production. Simple commodity production is typically built on smaller-scale, kin- or friendship-based work groups where there is widespread sharing of risks, costs, and benefits among owners and workers and significant variability in production. Simple commodity producers tend to be highly adaptive. As economic opportunities change over time, they can quickly take advantage of new possibilities.

In Qeqertarsuaq, the predominant form of simple commodity pro-

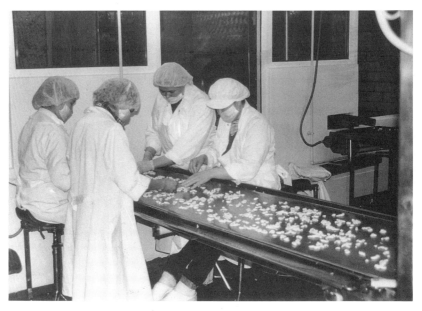

PLATE 2. Women working in Royal Greenland's shrimp processing plant

duction is shrimping. In 1993, Royal Greenland purchased 1484 metric tons of shrimp in Qeqertarsuaq, worth nearly $2 million (10.3 million DKK). In one year, a single vessel alone delivered shrimp worth over $500,000 (about 3 million DKK). The annual gross income of shrimp vessels delivering in Qeqertarsuaq averages about $26,000 (just over 143,500 DKK). In 1993, shrimp prices averaged about $1.31 (7.2 DKK) per kilo for premium-quality iced product. Four households in the municipality own or have major interests in ocean-going shrimp trawlers. These vessels range from 95 to 148 feet in length. Typically, they are manned by family or extended family members, though they often have nonfamily crew on board as well. Several other households own smaller shrimping or fishing vessels (30–60 feet in length).

In the 1980s and early 1990s, Royal Greenland's plant also purchased Atlantic salmon. Over 5000 kilos have been purchased in some years, valued at about $25,000 (140,000 DKK). In 1994 and 1995, however, Greenlandic commercial fishermen sold their entire salmon quota to a salmon conservation fund to help rehabilitate stocks. As a result, there was no commercial season in those years, although in 1996 fishermen once again fished salmon commercially. Other commercial species of local importance are lumpfish and crab.

In addition to fisheries, local residents also earn cash by selling hunting products: chiefly from seals, beluga, narwhal, and marine birds. Hunters sell these products through three different channels: (1) Royal Greenland's processing facility; (2) the local *kalaaliaraq* (English, "place where Greenlandic foods are sold"), an outdoor kiosk where fresh country foods are sold; and (3) private sales to local institutions (e.g., school or senior citizens' home) and to other households. Although Royal Greenland's primary focus is on shrimp production, it also buys other fishing and hunting products. Hunters get a better price from private sales, however, and Royal Greenland is usually the hunter's buyer of last resort.

Hunters also sell sealskins (primarily from ringed and harp seals) to Royal Greenland. In 1993, hunters sold 1680 skins locally. Prime-quality ringed sealskins normally bring about $41 (225 DKK), a price that includes a substantial Home Rule subsidy. Royal Greenland also purchases seal meat, as much as 8000 kilos in recent years. This meat is frozen and distributed for sale throughout Greenland. When they are available, Royal Greenland also buys beluga and narwhal products. *Mattak* from beluga and narwhal is a delicacy in Greenland, and there is great demand for it, particularly in the south. In one year recently, Royal Greenland purchased about 1600 kilos of beluga and narwhal meat and 1200 kilos of beluga and narwhal *mattak*.

The *kalaaliaraq* in Qeqertarsuaq is a small, open-air stall located near the harbor. The municipality provides the building as a clean and

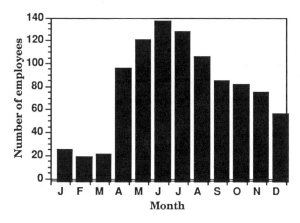

FIGURE 6. Seasonal nature of wage employment with Royal Greenland A/S, Qeqertarsuaq Municipality, 1988[11]

convenient place where fresh country foods can be sold. The *kalaali-araq* is used frequently in summer, when foods are typically available four or five days per week. In winter, it may be open only one or two days per week. Data from 1990 and 1991 reveal that, at Qeqertarsuaq's *kalaaliaraq*, fish was the most commonly available item, followed by seal meat, beluga or narwhal, birds, other whale meat, and plants.[12] Table 1 shows the types of foods typically sold and their prices. Prices are usually fixed annually by agreement between municipal authorities and the local hunters' and fishers' association.

A third channel for selling country foods is through private sales to institutions or to households. Hunters commonly sell directly to Qeqertarsuaq's school and senior citizens' home or to private households. The extent of these sales is difficult to gauge because income earned may not always be reported.

Household production of country foods

Most households in Qeqertarsuaq Municipality—including those with substantial cash incomes—procure at least some of their own country foods. Fully 90 percent reported doing so in Qeqertarsuaq, and 100 percent did so in Kangerluk.[14] Typically, men do the hunting, while women pick berries, process sealskins, and prepare foods. Not surprisingly, household members procure the most country foods during summer. The diversity of country foods obtained is striking. Table 2 shows the types of foods typically taken and the average amounts reported for calendar year 1989. Seals figure prominently in household catches, as

TABLE 1.
*Prices (in US$) for selected country foods sold in Qeqertarsuaq Municipality,
1993–1994*[13]

Product	Price (US$)/kilogram (1 US$ = 5.5 DKK)
Ringed seal meat (caught w/rifle)	7.26
Ringed seal meat (caught w/net)	5.65
Beluga/narwhal flipper or tail	9.68
Minke whale meat	5.65
Cod (dried whole)	8.07
Halibut	6.45
Capelin (dried)	11.29
Eider duck	6.45/each
Ptarmigan	5.66/each

TABLE 2.
Reported household catches in 1989 of selected country foods, by community[15]

Resource	Qeqertarsuaq (n = 56)		Kangerluk (n = 6)	
	Mean	Range	Mean	Range
Ringed seal	14	0–100	41	0–75
Harp seal	9	0–60	32	0–80
Beluga/narwhal	5	0–30	2	0–4
Walrus	<1	0–2	2	0–8
Hooded seal	0	0	2	0–10
Reindeer (caribou)	2	0–21	0	0
Ptarmigan	17	0–150	50	0–150
Eider duck	26	0–200	50	0–150
Murre	26	0–150	20	0–50
Kittiwakes	14	0–100	2	0–10
Atlantic cod (kg)	125	0–1000	442	0–1000
Atlantic halibut (kg)	13	0–125	49	0–250
Greenland halibut (kg)	3	0–100	33	0–200
Capelin (kg)	158	0–1500	1683	0–8000
Arctic char (kg)	49	0–450	42	0–150
Atlantic salmon (kg)	12	0–400	0	0
Greenland cod (kg)	136	0–1000	616	0–2000
Shrimp (kg)	0	0	0	0

do fish species. Beluga and narwhal also provide significant amounts of meat and *mattak*, especially in Qeqertarsuaq. Households in the two communities often have access to different species: Beluga and narwhal are more common in Qeqertarsuaq, whereas Kangerluk hunters have access to walrus. Kangerluk households also take significantly more capelin for dog food, reflecting the larger number of sled dogs there. Significantly, households in neither community catch shrimp for their own use.

By any standard, household subsistence production is high in both Qeqertarsuaq and Kangerluk (table 3).[16] Per capita production in Kangerluk, however, is twice that in Qeqertarsuaq; the levels are 1402 and 734 kilograms per capita, respectively. The higher levels of production in Kangerluk can be attributed to the fact that fewer income-producing opportunities exist there. If we subtract products used primarily for sled dogs, the average per capita production in both communities amounts to about 121 kilograms, a figure roughly comparable to Kapel and Petersen's finding that per capita consumption in hunting settlements averages about 150 kilograms.[17] Given these high levels of production, it is no surprise that consumption of country foods is also high. Nearly three-fourths of all households (73 percent) obtain most

TABLE 3.

Production of country foods by households in Qeqertarsuaq and Kangerluk

Community	Kg per household	Mean kilos per household member
Qeqertarsuaq (n = 56)		
Mean	2878	734
Range	5–12,755	
Kangerluk (n = 6)		
Mean	5370	1402
Range	140–11,572	

or all of their household meat and fish from wild foods. A majority of households eat wild meat or fish five or more days a week; 40 percent do so every day.

Cash plays an essential role in supporting country food production. Table 4 shows the purchase price of equipment typically used in these activities.[18] Capital investment can total over $17,000. The basic outfit for a hunter in the area includes a fiberglass skiff with outboard motor and a dog sledge and team of dogs. Most households own at least one skiff and motor, and a majority also own dogs. In Qeqertarsuaq, the average number of dogs per household is six, whereas in Kangerluk it is fifteen. Beyond these capital outlays, cash is also needed to operate and maintain this equipment in a demanding work environment. Gasoline alone costs about $.73 (5 DKK) per liter.

Integrating cash and subsistence: Household production strategies in Qeqertarsuaq and Kangerluk

A household survey conducted in 1990 reveals how local households integrate cash and country food production.[19] Survey results show that there are five basic household income strategies in the two communities (figure 7). The households differ significantly in their demographic composition, ownership of major capital assets (e.g., fishing vessels), income levels, participation in wage employment, and country food production (table 5). The first household category (category A) has only wage income; for the most part, these households are Danish, in Greenland for a short time. They have no income from simple commodity production and little or no subsistence production. They make up only 11 percent of all households in the two communities.

Households in category B are those owning shrimp trawlers or other large vessels. Although they make up only 13 percent of total households, they have the highest incomes and control significant capital assets. Their incomes are 28 percent higher than households without a vessel and 75 percent higher than those with limited incomes. Not surprisingly, they have significant loan payment obligations. They also typically own more hunting equipment than other households and produce significant amounts of country foods for household consumption.

Category C households do not own fishing vessels but, nevertheless, are active producers of fish and hunting products. They make up 29 percent of all households, including most in Kangerluk. Category C households produce the largest amounts of country foods per capita

TABLE 4.
*Costs of equipment typically used for subsistence production,
Qeqertarsuaq Municipality 1990*

Type of equipment/supplies	US dollars (1990 data)
Fiberglass skiff (16′ Pocco 500)	6400.00
Outboard motor (40hp Mariner)	4100.00
Fuel tanks and hoses (×2)	460.00
Shotgun (12 gauge)	450.00
Rifle (7.62mm Remington)	675.00
Rifle (.222 Sako)	890.00
Boat radio, battery, antenna	600.00
Plastic floats (×5)	300.00
Fish net/Arctic char (×2)	160.00
Fish net/salmon (×2)	110.00
Seal net (×4)	80.00
Binoculars	170.00
Dog sledge	470.00
Dog harnesses and lines (×9)	200.00
Dog sledge pad (caribou hide, ×2)	65.00
Sled dog whip	75.00
Harpoon shaft and head (×2)	450.00
Ice chisel	29.00
Tent	360.00
Sleeping bag	130.00
Survival kit with flares	110.00
Fiberglass punt (8.5′)	836.00
Campstove and tank	38.00
Walkie-talkie	282.00
Ammunition (12 ga./25 shells, ×5)	40.00
Ammunition (7.62mm/15 shells, ×5)	45.00
Ammunition (.222/20 shells, ×5)	32.00
Gasoline (5 DKK/liter, ×40 liters)	36.00
Total	$17,593

SOURCE: Field data.

PLATE 3. Hunter and dog sledge on the sea ice, Disko Bay

- Household Category A: wage income only
- Household Category B: simple commodity production; fishing vessel owners
- Household Category C: simple commodity production; no fishing vessel
- Household Category D: mixed wage and subsistence production
- Household Category E: limited income (e.g., pensions)

FIGURE 7. Household categories in Qeqertarsuaq Municipality
(adapted from Caulfield 1991)

of all households, and most of their income comes from selling these products. They are also the highest household consumers of country foods. Unlike vessel-owning households, they have fewer loan obligations.

The largest proportion of households (39 percent) falls into category D—those with mixed wage and country food production. Wage income in these households is generally higher; usually one or more adults works full-time for wages. Although there is little simple com-

TABLE 5.

Selected household characteristics, by household category, Qeqertarsuaq Municipality

Household category	Mean household size	Percent Greenlandic speaking[a]	Mean number of vessels owned	Mean number of skiffs	Mean number of sled dogs	Mean production country food[b] (kg/person)	Mean household income (000s DKK)	Mean value means of production[c] (000s DKK)
A: Wage income only	3.14	42.90	0.00	0.14	0.43	1.65	286.70	15.00
B: Simple commodity production/vessel owner	4.75	87.50	2.00	2.25	9.00	990.63	446.90	6253.00
C: Simple commodity production/no vessel	4.39	94.40	0.00	1.89	9.72	1275.07	321.50	148.30
D: Mixed wage & subsistence production	3.79	100.00	0.00	1.13	6.08	511.69	279.30	68.90
E: Limited income	2.60	100.00	0.00	0.80	1.20	442.89	110.00	54.2

SOURCE: Adapted from Caulfield 1991.
[a] Households speaking exclusively Greenlandic.
[b] Country foods (including sled dog food) procured for household consumption.
[c] Estimated present value of equipment typically used to procure country foods (including fishing vessels).

modity production in these households, they consume considerable amounts of country foods. Cash earned in wage employment is typically used to purchase country foods that household members themselves are not able to obtain.

The final household category (category E) is made up of those with limited incomes, especially pensions. They have no income from wages or simple commodity production. These households make up 8 percent of the total surveyed. Their household size is smaller, reflecting the fact that few children are living in the home. Surprisingly, they produce considerable amounts of country foods for their own use. Most of this production is fish taken in summer months. These catches likely reflect the time these household members have to invest in country food production and the importance of country foods to elders.

The data reveal that households in Qeqertarsuaq employ different economic strategies than do those in the smaller settlement of Kangerluk. The latter have lower wage incomes and higher country food production (table 6). Consistent with this result, they also typically own more hunting and fishing equipment and sled dogs. These data are consistent with findings in other studies showing how settlement households are more reliant on country foods, have lower incomes, and own more of their own means of production than those in larger towns.[20]

Kinship, sharing, and the ideology of subsistence

Greenlanders, like other Inuit, are recognized as having generally egalitarian economic systems where all members have access rights to resources. According to Guemple, Inuit social organization is

equalitarian rather than hierarchically structured, inclusional (that is, calculated to pull people in) rather than exclusional, organized on proximity (and therefore variable) rather than along genealogical lines (and therefore rigid), and based on a principle of generalized exchange rather than on highly restrained and elaborately structured forms of cooperation.[21]

The core of these systems is kinship, as Eric Wolf emphasizes in his definition of a kin-ordered mode of production.[22] In this mode, production groups usually consist of primary relations of lineal kin or close affinal marriage relations. Flexibility is equally important, however. Members can create partnerships and cross-cutting alliances when the need arises. As Wolfe and colleagues note,

TABLE 6.
Comparison of selected household data for Qeqertarsuaq and Kangerluk

Variable	Qeqertarsuaq	Kangerluk
Mean hh size (# persons)	3.93	3.80
% Greenlandic speaking	89.30	100.00
Annual subsistence production, kg/capita	732.34	1413.38
Mean # sled dogs in hh	5.57	15.00
Mean # skiffs in hh	1.32	1.67
Mean # trawlers	0.89	0.00
Mean value of MOP* (000s DKK)†	954.77	209.50
Mean # hh members earning wages, 1989	1.93	1.00
Mean hh income (000s DKK)	312.49	219.00
Mean hh income from fishing/hunting (000 DKK)	14.12	84.80

SOURCE: Caulfield 1991.
  * Means of production.
  † Includes fishing vessels.

it is a system which conceives an autonomy of action in subsistence pursuits by familial groups, where capital and labor are held and controlled autonomously by relatively small-scale kin groups. . . .

[I]t is a system which is geared to produce for finite objectives—the maintenance and continuance of the local sociocultural system.[23]

The primary goal of this system is mutual security, based upon collective responsibility in production activities and sharing of subsistence products.[24]

In Qeqertarsuaq and Kangerluk, kinship is the chief organizing principle for both simple commodity production and for procuring country foods. Work groups for beluga hunting, char fishing, or berry picking are almost always made up of immediate or close extended family members (*ilaqutariit*). Figure 8 shows two examples of such production units: one for capelin fishing and the other for beluga hunting. But people are also flexible in forming work groups, and group composition can change according to the season and resource involved. Short-term alliances or partnerships are common. Under this flexible arrangement, hunting products can be shared more widely, as when distributing the proceeds of a whale or walrus hunt. Partnerships, furthermore, enable older people to share their knowledge and experience with those who are younger. Even in larger-scale production activities—working on shrimp trawlers, for example—kinship ties are often present. One of Qeqertarsuaq's shrimp trawlers is owned by five brothers, all of whom have a different role in shrimp production.

In Qeqertarsuaq, country foods provide a nutritious, high-calorie diet that is well suited for the Arctic (figure 9). A survey of meat and fish consumed by selected households in Qeqertarsuaq in 1989 and

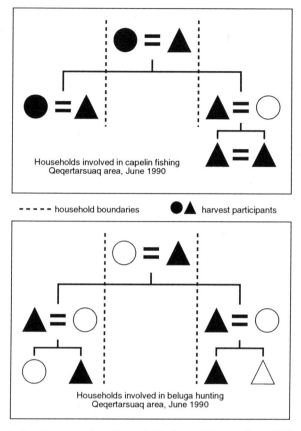

FIGURE 8. Social organization of households for catching capelin and beluga near
Qeqertarsuaq, 1990

1990 reveals that fish was the most frequent food consumed (23 percent), followed by seal and walrus meat and imported products. Although households today have access to a wide variety of foods, most
consider marine mammal products to provide the best nutrition. Seal
meat, in particular, is viewed as "real food."

The consumption of country foods is closely interwoven with systems of kinship and sharing. In West Greenlandic, country foods are
called *kalaalimerngit*, or Greenlanders' foods. Greenlanders differenti-
ate these foods from Danish or other imported foods, called *qallunaam-
erngit* or white man's food. As survey data show, *kalaalimerngit* comprise
a substantial part of household diets; but these foods have significance

beyond simply their nutritional value. The processes of procuring, processing, preparing, and sharing Greenlandic foods help bind families
and communities together. Sharing foods reflects underlying systems
of reciprocity and community solidarity that continue to give meaning
to people's lives. They are, moreover, important markers of Greenlandic identity. As Larsen and Hansen point out, "this distinction between 'Danish food' and 'Greenlandic food' is far more significant than
a merely functional distinction referring to the origin of the food.
Eating Greenlandic food is of great symbolic weight in determining
whether a person is a true Greenlander."[25] The rich diversity of country
foods available provides a sense of security for local people that is difficult for non-Greenlanders to appreciate. Though ships arrive regularly
bringing imported foods, one can sense a special satisfaction among
families as they share *kalaaliminertorneq*, a sort of smorgasbord of wild
Greenlandic foods.

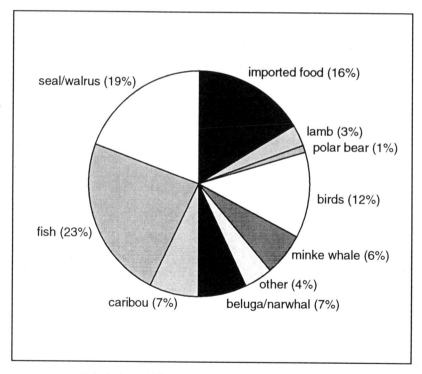

FIGURE 9. Principal meat/fish consumed by selected households in Qeqertarsuaq,
1989–1990 (source: field data)

. . . our host invited us to share *kalaalimerngit* with him and his family. We relaxed and drank tea while his young son went down to their outdoor cache to fetch the frozen foods. There was a flurry of activity in the kitchen, and shortly after we were invited to sit around a table literally covered with a rich variety of Greenlandic foods. There was *tuttu panertut* (dried caribou meat), *saarulliit panertut* (dried cod), *puisip tingua* (frozen seal liver), and *angmassat panertut* (dried whole capelin). We also had *tikaagulliip qiporaa* (belly flesh from minke whale), *quaq* (frozen seal meat), *tikaagulliip sarpia tarajugaq* (mattak from minke whale tail), *qilalukkap qaqortap mattaa* and *tinguanik imerlugu* (beluga mattak and frozen liver), and *tuttup iloqutai* (caribou mesentary fat).

We tasted everything, tearing off pieces of dried meat or fish with our hands, or cutting off pieces of mattak with a sharp knife. We placed these pieces on a small wooden board in front of each of us, and ate small slivers of each food as our hosts quietly did the same.[26]

People in Qeqertarsuaq love to share country foods outdoors with their families on summer days. Entire extended families—from babies to elders—gather in sheltered locations on the outskirts of town to cook over a small fire made of crowberry and dwarf birch. These foods are also enjoyed at a distinctively Greenlandic celebration called *kaffemik*. These ceremonies are held to celebrate birthdays, anniversaries, baptisms, and confirmations. At a *kaffemik*, guests simply stop by at any time during the afternoon or early evening. After entering the home and giving special greetings to the honored person and family members, guests sit at a table brightly decorated with candles and brimming with cakes and sweets. Many of the foods served are of Danish origin, but depending upon who the guests are, local foods like beluga *mattak* and dried fish may also be served. Guests may find a bowl containing small pieces of caribou fat to place in their tea, making a delicious broth. Special meals of *kalaalimerngit* are also served upon the arrival of guests from other communities or at community-wide celebrations such as Greenland's national day on June 21.

Petersen describes how, in Greenland, ideologies associated with sharing country foods are changing over time.[27] Until recently, households typically shared on a generalized basis, which fostered solidarity and provided a sort of insurance against difficult times. The meat-gift system also helped support those who were not able to hunt for themselves. As settlements grew larger, these patterns began to change. Meat-gifts were increasingly restricted to close relatives and neighbors.[28] Changes became particularly pronounced with the introduction of capital-intensive technology (like fishing vessels). More products were sold to get the cash necessary for operating and maintaining this equipment.

Despite these changes, the sharing of food remains an important part of everyday life in Qeqertarsuaq and in Kangerluk. Half of all households report that they often or always share country foods with others. The most common recipients of shared foods are immediate family (98 percent of the time) and extended family (78 percent). Friends and acquaintances receive country foods about 40 percent of the time. Foods are also sent to family members living in other communities.

Hunters distribute gifts of meat or fish (called *pajugat*) under a variety of circumstances. Gifts might simply be made to those whom the hunter likes or to express appreciation to those who have helped in some way (*qujagisaqarneq*). Country foods are also given to those sharing names (*atsiaqarneq*).[29] The named person, or *atsiaq*, will often receive gifts from the family of the person for whom he/she is named. The *atsiaq* will refer to the father in the other household as his *ataatakulooq*, and to the mother as *anaanakulooq*, and will commonly spend time with that family. Hunters might also share meat or fish with those unable to hunt for themselves (*pilersuisoqanngitsut*) or to those lacking the means to hunt or fish for themselves (*piniutoqanngitsut*).

The types of meat-gifts frequently differ depending upon the gender of the recipient. For example, boys usually receive the *puisip tajarneq*, or the front flipper and claws of a seal. This gift is said to give the boy strength in hunting. In contrast, girls receive the *puisip pamialluk*, or the lower vertebrae and coccyx (tail bone) of a seal. Significantly, gifts of meat and fish are made without consideration of the wealth of the recipient household. The owner of a shrimp trawler, a highly paid wage employee, and a self-employed fisherman can all be equal participants in sharing. The significance of sharing thus extends beyond simply provisioning households; it serves to reinforce the collective solidarity of families and communities.

Continuity and change in Qeqertarsuaq Municipality

The common theme in this account of production relations in Qeqertarsuaq is change. Change occurred when the first Norse hunting parties appeared in Disko Bay nearly a millennium ago. It accelerated when Dutch and other European whalers began trading in the seventeenth century and when Hans Egede brought his Christian flock to Greenland to save souls. Shortly after, European whaling decimated stocks of bowhead whales that Greenlanders had used for generations. This difficulty was compounded when climatic changes forced a shift from seal hunting to fishing as a major occupation in much of West Greenland. Change is still evident today, as people respond to the dynamics of shrimp, cod, and salmon stocks.

Anthropologist Jens Dahl argues that flexibility and an ability to respond appropriately to change may well be hallmarks of Greenlandic culture.[30] This feature certainly seems evident in Disko Bay. Again and again, Greenlanders in Qeqertarsuaq and Kangerluk have revitalized and reinvented ancient traditions in response to new realities. Wage employment is a part of those realities today, and often hunting and fishing must be scheduled around work hours. Over two-thirds of the households in the municipality say that they spend less time hunting now than they did ten years ago;[31] however, new technologies now enable hunters to be more efficient in procuring foods. Sharing of country foods between households is also changing as more and more hunting products are sold to obtain cash for heating oil, outboard motors, and imported products. In Qeqertarsuaq, over three-quarters of all households share less now than they did twenty years ago. For some, at least, sharing country foods is becoming more a symbol of shared ties than an actual provisioning system.

The impact of world political and economic systems is a part of these changing realities. Household economies in Greenland are affected to a large extent by world markets for shrimp, Home Rule development policies, and Danish attitudes toward continued subsidies. In recent years, prices for Greenlandic commodities have suffered from declining terms of trade on world markets. Prices of everything from shrimp to sealskins to salmon have fallen, and the burden is shared by all Greenlanders. For example, world prices for unprocessed shrimp declined by 15 percent between 1987 and 1992, squeezing fishermen and processors at a time when production prices were increasing. The salmon fishery is plagued by biological concerns and flagging prices due to competition from farmed fish; and sealskin prices are strongly influenced by global politics, especially decisions made by the European Community banning sealskin imports.

As Greenland moves toward greater privatization, new concerns arise about social differentiation, not only between towns and settlements but also within communities. Even now one hears references to the so-called "millionaires" of the shrimp industry—those (especially in Disko Bay) who supposedly have profited from developments in the shrimp fishery. To some extent this differentiation may reflect long-standing stratification evident even within "egalitarian" Inuit society, where prestige, power, and privilege accrued to the *piniartorsuaq*, or "great hunter," who could provide for many. But does the expansion of simple commodity production and wealth in general in Greenland necessarily imply development of an ever-more stratified social system?

In Qeqertarsuaq Municipality, the shrimp fishery appears to be a major factor contributing to social differentiation. As the previous data show, households involved in this fishery control far more assets than do others. Given that lucrative shrimp quotas are now allocated by the state and are transferable, capital might continue to accrue to those involved in this fishery.

Wolfe and colleagues, however, have found that leveling mechanisms serve to counter tendencies toward social stratification in Alaskan Inuit communities involved with commercial salmon fishing.[32] Evidence there suggests that stratification was limited to the salmon fishery and did not extend to other areas of household economies. Researchers identified four leveling mechanisms: (1) egalitarian property rules, where rights to resources were shared widely; (2) persistence of sharing practices between households; (3) bilateral inheritance, reducing consolidation of inherited resources in the hands of a few; and (4) the variety and mobility of wild resources, limiting the ability of any one social group to dominate catch efforts.

In Qeqertarsuaq Municipality, similar economic leveling mechanisms appear to be at work. At the household level, property rules are flexible, allowing frequent sharing of personal property such as boats, nets, rifles, and so forth. Even personal wage income is often used to buy equipment benefiting the entire household. Likewise, sharing between households is both an economic leveler and a source of prestige in Greenlandic society. A provider—whether an elder hunter or a young wage-earning fisherman—is respected for his generosity and for contributing to the mutual security of the community. Rules of inheritance also mitigate against inordinate accumulations of wealth. In most cases, items from the estate of a deceased family member (e.g., a boat or rifle) continue to be used for the benefit of the extended family. Preference is frequently shown toward one survivor (especially an eldest son) in distributing these items, though today there are signs that Danish legal practices are changing this practice. Finally, the variety and mobility of marine resources generally mitigate against any one segment of a community gaining advantage over another.

These concerns about social differentiation—especially between those in towns and settlements—are a continuing element of political discourse in Greenland. As we shall see in Part Two, they are exacerbated through Greenland's whaling regime, where tensions have increased because of external forces and involvement in co-management.

PART TWO

LOCAL DYNAMICS, GLOBAL
CONCERNS: CONFLICT
AND CONTRADICTIONS
IN GREENLANDIC WHALING

Across the world, grassroots movements are working to open up more space for the commons by denying that any single whole — whether culture, language, livelihood, art, theory, science, gender, race or class — has a right to assert privileged status over, and thus to enclose, all others of its type. . . . Key to that struggle is the building up of open and accountable institutions that restore authority to commons regimes — a struggle which requires increasing the bargaining power of those who are currently excluded or marginalized from the political process and eroding the power of those who are currently able to impose their will on others. Only in this way . . . can the checks and balances on power that are so critical to the workings of the commons be ensured.[1]

≋ 3 ≋

Vessels, Kin, and Harpoons: Continuity and Change in Whaling

Introduction

P art One of this book describes Greenland's growing interaction with global political and economic systems and illustrates this contact with the case study from Qeqertarsuaq. Part Two focuses on Greenland's aboriginal subsistence whaling regime for minke and fin whales under IWC guidelines. In it, I describe changes in whaling practices and management over time and the institutions and processes that make up the contemporary Greenlandic regime. I evaluate this regime from a co-management perspective, focusing in particular on transaction costs Greenlanders incur from their involvement with it. The final chapters highlight the importance of self-determination as a foundation for sustainable aboriginal whaling. I discuss how the Home Rule government is seeking to overcome marginalization in the IWC through alternative political strategies that appear to reduce transaction costs and thus strengthen local control over whaling.

Whales in Greenlandic waters

More than ten species of whales and dolphins inhabit the waters off Greenland, including Davis Strait, Baffin Bay, and Denmark Strait (table 7). In recent history, Greenlanders have taken bowhead, minke, fin, and humpback whales, along with beluga, narwhal, pilot whales, and porpoises. Intensive European exploitation of bowheads for over two hundred years drove the Davis Strait stock almost to extinction. Since the end of commercial bowhead whaling in the early twentieth century, the stock has been slow to recover. Biologists believe that bow-

TABLE 7.

Names of whales and dolphins commonly used by Greenlanders during the historic period

English name	Latin name	West Greenlandic name
Bowhead or Greenland right whale	*Balaena mysticetus*	*arfivik*
Minke whale	*Balaenoptera acutorostrata*	*tikaagullik*
Fin whale	*Balaenoptera physalus*	*tikaagulliusaaq*
Humpback whale	*Megaptera novaeangliae*	*qipoqqaq*
Beluga or white whale	*Delphinapterus leucas*	*qilalugaq qaqortaq*
Narwhal	*Monodon monoceros*	*qilalugaq qernertaq*
Blue whale	*Balaenoptera musculus*	*tunnulissuaq*
Sperm whale	*Physeter macrocephalus*	*kigutilissuaq*
Sei whale	*Balaenoptera borealis*	*tikaagulliusaarnaq*
Killer whale	*Orcinus orca*	*aarluk*
Common porpoise	*Delphinus delphis*	*aarluarsuk*
Harbor porpoise	*Phocoena phocoena*	*niisa*
Pilot whale	*Globicephala melas*	*niisarnaq*

heads in Davis Strait and Hudson Bay now number at least 450 animals, but this count is still less than 5 percent of the estimated initial stock size of about 12,000.[2] The IWC considers these whales to be a "protection stock," and no catches are allowed.

Today, the two large whales taken in Greenland under IWC quotas are the minke and the fin whale. Greenlanders took humpback whales as recently as 1985, when the quota was set at zero. Both minkes and fins, like other baleen whales in the North Atlantic, follow a seasonal migration pattern into northern latitudes in summer and back to warmer waters in fall and winter.[3] The minke is the smallest species of baleen whale, typically measuring about 10.5 meters in length. Its blow is low and inconspicuous, making it difficult to see from a distance. It is a fast swimmer and is often found singly or in pairs or trios. It also appears to segregate by age and sex more than other baleen whales. Minkes eat fish (e.g., capelin, Greenlandic cod, sand eel) and invertebrates.

Minkes are found in both East and West Greenland.[4] Hunters in southern West Greenland usually begin catching them in April and continue to do so through November.[5] The whales are also occasionally killed during winter months. Catch records in Greenland suggest a peak abundance of minkes off West Greenland in early summer (May) and in late autumn (October). The IWC Scientific Committee estimates that the West Greenlandic "stock" of minke whales numbers

PLATE 4. Fin whales off Qeqertarsuaq

some 8371 whales (95 percent CI: 2,414–16,929), although there is agreement that West Greenlandic minkes are very likely part of a larger, but as yet undefined, stock.[6] Population estimates for the central Atlantic stock of minke whales—including those near East Greenland—indicate that there are about 28,000 whales (95 percent CI: 21,600–31,400).[7] The IWC categorizes whales in both of these areas as "protection stocks." This designation means that they are below levels allowable for any commercial whaling (though it does not necessarily mean that they are nearing extinction).[8]

Like minkes, fin whales are also found in both East and West Greenland during the summer months. Little is known about their wintering areas, although they are found as far south as Florida and the Gulf of Mexico. Fin whales are much larger than minkes, measuring up to twenty-four meters. They eat krill and other invertebrates as well as fish (e.g., capelin, sand eels). Fin whales are most often found in pods of three or more animals and are known for their speed. Their blow is powerful and tall (four to six meters), making them easy to spot from long distances. In Greenland, their numbers peak during northerly migration in June and July and again as they move south in September and October.[9] At times, they appear to congregate where food supplies are abundant. For example, in Qeqertarsuaq in 1989, over twenty fin whales were observed feeding in a small area. Hunters there catch them from June through October, with peak catches in September. Estimates of fin whale numbers near West Greenland suggest a population of about 1046 (95 percent CI: 520–2,106).[10] Recent efforts to determine if fin whales in West Greenland are distinct from those elsewhere in the North Atlantic were inconclusive, with no significant differences found.[11]

Humpback whales can measure up to sixteen meters in length. They tend to inhabit shallower coastal areas, eating krill and schooling fish. They make long seasonal migrations between summering areas in the North Atlantic and wintering areas to the south. Biologists believe there are at least four and perhaps five separate feeding substocks in the North Atlantic, including one in West Greenland and one in the Iceland-Denmark Strait area. Humpbacks are rather slow swimmers, traveling between six and twelve kilometers per hour. This fact, and their pattern of sleeping on the surface, makes them easier to catch than other whales. Recent estimates of the total humpback whale population in the North Atlantic are 5505 ± 2617 (95 percent CI).[12] Humpbacks feeding near Greenland are believed to number about 430 ± 151 (95 percent CI).[13] The IWC considers all North Atlantic humpbacks to be included within a "protection stock" and allows no hunting.

Greenlandic hunters are keen observers of whales. Older hunters, in particular, often have extensive knowledge of whale behavior, feeding habits, and seasonal distribution. Names for whale anatomy and edible parts, shown in table 8, are well represented in the West Greenlandic language. In addition, oral traditions relate considerable information about whale behavior and distribution. For example, Greenlanders are aware that the most productive whaling areas over time correspond with the location of deep underwater trenches lying between major fishing banks, particularly near Paamiut, Qaqortoq, and Disko Bay.[14] This knowledge is augmented with awareness of historical changes in marine and terrestrial resources distribution and abundance.

Historical overview of Greenlandic whaling

Figure 10 depicts major eras of Greenlandic whaling, from prehistoric times to the twentieth century. From the earliest times, Greenlanders have survived in a marginal Arctic environment by accommodating themselves to changing ecological, political-economic, and historical factors. Whaling practices are no exception. Resource availability, technologies, and social systems have all changed over time. As a result, there is no static, timeless, or "pure" Greenlandic whaling; change is a constant. Older practices continually give way to new traditions as needs arise.

Archaeologists' discoveries of whale bones, baleen, and teeth dating to 2400 B.C. in southern Disko Bay confirm that early Greenlanders made extensive use of whales.[15] At this Disko Bay site, generations of Greenlanders used baleen from right whales, bones from either minke or sei whales and from killer whales, teeth from sperm whales, and narwhal tusks. It isn't clear whether these products came from hunted whales or were scavenged from carcasses washed up on beaches; however, clearly whales were available during this period, and they played an important role in the daily lives of early Greenlanders.

Whaling was one of the distinctive elements of Thule (Inugsuk) culture, which developed in Greenland about A.D. 1000 to 1100. Thule Inuit brought with them tools used in whaling, including the *umiaq* and distinctive whaling harpoons. Hunters focused particularly on bowhead and humpback whales because of their slow speed and habit of sleeping on the surface. These characteristics made them easier to approach and kill.[16] Little is known about Thule whaling practices, but clearly the *umiaq* was an essential piece of equipment. It was paddled by six to eight hunters, with a person steering in the stern and one or

TABLE 8.
West Greenlandic names for whale anatomy

English name	West Greenlandic name
Blowhole	*kiingaq*
Back fin	*naparutaq*
Region between body and tail	*singerneq*
Fluke	*sarpik*
Flipper	*taleroq*
Fluted area below mouth	*qiporaq*
Mouth	*qaneq*
Body	*timaa*
Baleen	*soqqaq*
Ear bone	*siunnak*
Skin (muktuk)	*mattak*
Blubber	*orsoq*
Meat	*neqi*
Baleen whale jawbones	*alleruit*

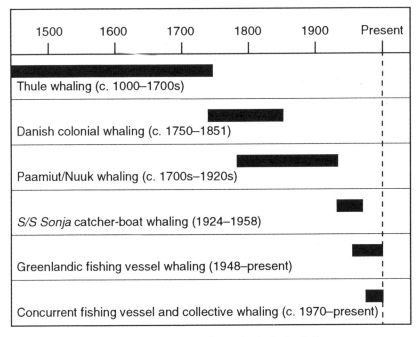

FIGURE 10. Recent eras of Greenlandic Inuit whaling

two harpooners in the bow.[17] Floats were installed under the seats to keep the craft afloat if it capsized. Hunters in several *umiat* would work together to affix as many harpoon heads as possible in the whale. The detachable harpoon heads, measuring about twenty centimeters in length, were made of whale jaw bone and fitted with a stone or iron blade. The hunting line was probably made of bearded seal or walrus hide, about thirty to thirty-five meters or more in length. Several floats made of bearded seal hide were attached to this line.[18] After a whale was mortally wounded, hunters would often use the *atallaaq*, a specialized "dry suit" used for crawling onto a whale in the water to deliver the final strike and to aid in flensing. Flensing took place both at sea and on shore.[19]

Whales had great spiritual significance for these Greenlandic hunters. The propitiation of animal souls was a central element in Inuit cosmology.[20] Whales and other sea mammals were the gifts of *Sassuma Arnaa*, the "woman-of-the-sea," who made them available to humans as long as proper taboos and rituals were observed. Right behavior was a responsibility not only of hunters but of the entire community. When men went hunting, women had to remain indoors in darkness. These beliefs persisted in the early colonial period; one European observer noted that ". . . when they sail out for whale fishing they dress themselves up in their finest clothes ostensibly because the whale demands respect and no filth will tolerate."[21]

Early Greenlandic hunters observed property rights to whales that were rooted in customary law and practice. These practices governed the common ownership of harpooned whales, distribution of whale products, and appropriate behavior for those involved in whaling and flensing.[22] Because large whales were considered a common resource belonging to all, products from whales were shared widely. Blubber, meat, *mattak*, baleen, and teeth were all utilized. Baleen was particularly valued for use in fishing nets. Importantly, the enormous size of whales meant that even households without an active hunter (e.g., those comprised of widows and orphans) could obtain substantial quantities of both food and oil for soapstone lamps.[23]

Whaling in colonial Greenland

As described in chapter 1, regular contact between Greenlanders and Euro-Americans began in the seventeenth century when Dutch traders began visiting West Greenland. Intensive European whaling in Davis Strait began in the early 1700s. Early traders found a flourishing Inuit exchange economy extending the length of Greenland's west coast. The principal commodities exchanged were baleen from Disko Bay and furs and soapstone from southern Greenland. As one merchant reported in 1752,

in the South the Greenlanders themselves use most of the baleen, despite the fact that they have none. Therefore they have to go north to Disko Bay to get it. On their way north they supply the markets with their many fox furs, supposedly to the Dutch. In the Disko Bay the Greenlanders need caribou skin and soapstone pots which too are lacking in the South.[24]

Climatic changes during the Little Ice Age (about 1650 to 1750) limited Greenlanders' access to large whales and altered settlement and trading patterns. As European trade goods such as iron, fish hooks, and brass kettles became more common, the exchange system was also modified. Greenlanders from the south began making longer, multi-year journeys north to the Disko Bay region.[25]

Growing contacts with European whalers and the advent of Danish colonization in 1721 brought two major changes to Greenlandic whaling. First, Greenlandic hunters began adopting European whaling technology, including the use of harpoons with metal heads and more efficient flensing tools. By the late 1700s, hunters were using the European whaling sloop, or pinnace, instead of the *umiaq*.[26] The second change occurred when colonial whaling—where Greenlanders were hired as crew—became more common. In the process, Greenlandic customary law regarding disposition of the catch was increasingly influenced by economic considerations.[27]

Between 1750 and 1851, Danish authorities in Disko Bay, Sisimiut, and elsewhere along the coast attempted land-based whaling, first using European crews and later employing local hunters. In the 1790s, colonial whalers averaged twenty to thirty bowhead whales a year, usually catching them between January and March.[28] Local hunters sold the blubber to KGH and kept the whale meat for themselves. As bowheads became scarce in the early 1800s, however, colonial whaling became increasingly unprofitable. Catches in Disko Bay declined from over fifty between 1805 to 1809 to only a single whale annually in later decades.[29] Colonial whaling finally ceased in 1851, when the station at Qeqertarsuaq was shut down.[30]

European whaling vessels continued to visit Disko Bay and West Greenland in small numbers until the beginning of the twentieth century. Bowhead whales remained their primary focus, but they took other species as well. Bang notes that in 1896 Scottish whalers took six bowhead whales in Davis Strait/Baffin Bay and five near East Greenland.[31] Winge reports that one whaler took fifteen humpbacks in Disko Bay in 1868.[32]

Petersen describes the period following the end of colonial whaling as a time when Greenlanders lost a great measure of self-respect.[33] Within only a generation or two, foreign whalers had nearly destroyed bowhead stocks, undermined longstanding Inuit customs and beliefs, and introduced diseases that decimated entire communities. Despite this demoralized state, hunters in Disko Bay and elsewhere continued to catch whales sporadically through the end of the century.[34] For example, oral traditions in Qeqertarsuaq relate how *Piitarsuaq* (Peter Carl

Niels Broberg; b. 1825, d. 1902), a renowned hunter and forebear of families still active in whaling today, caught both bowhead and minke whales.[35] *Piitarsuaq* assisted a Scottish whaler visiting Disko Bay who, in appreciation, gave the Greenlander a small sloop. According to *Piitarsuaq*'s son, the hunter used this equipment to catch at least two bowheads (probably in 1882)[36] as well as a number of minke whales.

I got to go out with the whalers [in Qeqertarsuaq]. I went along with my father when he went whaling. I was getting to be a young man [13 years] when my father caught two *arferit* [Greenland right whales]. He lost a third one, which washed up in Attu [on the south side of Disko Bay]. . . . But we also caught a large number of *tikaagullit* [minke whales].[37]

In Qeqertarsuaq, Greenlanders' interest in catching bowheads continued into the twentieth century.[38] An unsuccessful attempt was made in the late 1920s, and one whale was successfully taken in 1972. Not infrequently, hunters also found dead whales on nearby beaches and used them for sled dog food.[39]

Greenlandic whaling in the twentieth century

Greenlandic catches of large whales during the twentieth century occurred largely within four successive eras: (1) humpback whaling in Paamiut and Nuuk using small boats, continuing into the 1920s; (2) Danish catcher-boat whaling between 1924 and 1958 using the vessels S/S *Sonja* and S/S *Sonja Kaligtoq*; (3) fishing vessel whaling with harpoon cannons, begun in 1948; and (4) concurrent whaling by both fishing vessels and a collective hunt involving small skiffs and outboard motors, beginning in about 1970.[40]

Humpback whaling continued near Paamiut and Nuuk until the mid-1920s, with as many as twenty-two caught annually.[41] As late as 1928, whalers in Paamiut attempted to catch humpbacks.[42] Greenlanders in Nuuk, Maniitsoq, and Aasiaat also took humpbacks during this period;[43] however, the total annual catch in Paamiut and Nuuk averaged fewer than five animals (figure 11).[44] Hunters in these communities originally used *umiat* but later switched to European equipment. As in earlier *umiaq* hunting, hunters often approached and harpooned humpbacks while they were sleeping on the surface, a type of hunting known as *pussinnat*.[45] The persistence of whaling during this period

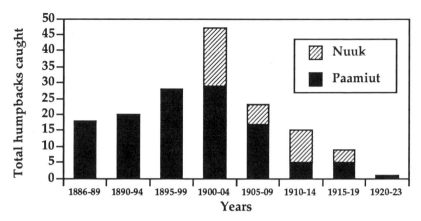

FIGURE 11. Greenlandic catches of humpback whales in Paamiut and Nuuk, 1886–1923
(source: Kapel 1979)

suggests that it remained important to Greenlandic livelihoods and beliefs. Its continuation was due in part to the efforts of John Møller, a
Greenlandic photographer and hunter who sought to revitalize whaling
in local communities.

In 1912, a Danish captain, H. V. Bang, proposed to the Danish
"Greenland Association" (*Det grønlandske selskab*) in Copenhagen that
colonial authorities take up whaling once again to cope with food
shortages in West Greenland.[46] A Norwegian vessel had demonstrated
the viability of this idea with a small whaling operation in Davis Strait.
Although World War I postponed action on Bang's idea, the notion set
the stage for Danish colonial whaling over the next several decades.
In 1924, KGH purchased a Norwegian catcher-boat, the 127-ton M/S
Sonja. The *Sonja* was built in England in 1910 and was used in Norwegian Antarctic whaling until 1923.[47] The ship had a three-hundred-
horsepower, coal-fired steam engine and was capable of doing nine
knots. She was equipped with a bow-mounted harpoon cannon capable
of taking the largest whales.[48]

From 1925 to 1928, *Sonja* caught whales off Greenland's west coast
and delivered them to a flensing vessel, M/S *Sværdfisken*. This ship,
which had also been purchased by KGH in 1924, was outfitted with a
steam winch and rendering equipment on deck.[49] Beginning in 1929,
the *Sværdfisken* was no longer used for flensing. Instead, *Sonja* delivered
whales directly to local communities where residents flensed the

whales, receiving meat and innards for their efforts. To pay for *Sonja*'s operation, whale blubber was shipped to Copenhagen for rendering and sale by auction.

Sonja's first year in Greenland resulted in twenty-nine whales being caught (figure 12).[50] The principal species taken were blue, fin, humpback, and sperm whales. The majority of the kills were delivered to communities in southern Greenland, which had suffered declines in seal catches. *Sonja*'s catches were also important, however, in northern West Greenland, where whale products were valued for sled dog food. *Sonja*'s approach with a whale generated great excitement in a community. According to Smidt,

there was one interruption in everyday life which was not unwelcome. That was when *Sonja* came in with a whale. . . . I especially remember an evening in July of 1948, when *Sonja* came into Uummannaq with an enormous blue whale, a male about 25 meters in length. It created quite a commotion in the colony when the whale was hauled in and the flensing began. Everyone was up and about, big and small, as long as the flensing lasted, and there was plentiful meat for all households both for people and dogs, besides the meat hung to dry on the racks. . . . *Sonja*'s popularity found expression in an especially catchy

PLATE 5. Danish whaling vessel M/S *Sonja*
(photo courtesy the Greenland National Museum, Nuuk, Greenland)

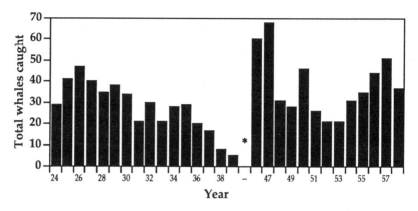

FIGURE 12. Catch of large whales by S/S *Sonja*, 1924–1950, and *Sonja Kaligtoq*, 1951–1958; * indicates no whaling during WWII (source Kapel 1979)

song known throughout the country, "*Sunija kaligpoq*," or "*Sonja* is towing a whale."[51]

Sonja's Danish captain and crew of thirteen caught an average of twenty-three fin whales annually during the pre-war era, although catches declined in the 1930s due to poor weather and competition from Norwegian whalers operating in Davis Strait.[52] Whaling was suspended during World War II, and *Sonja* was confined to drydock in Maniitsoq. Following the war, she was refitted in Denmark and resumed operations; however, it soon became clear that a newer and faster ship was needed.

In 1950, KGH purchased a replacement vessel in Norway, the 250-ton *Sonja Kaligtoq* ("the one that tows [whales]"). *Sonja Kaligtoq's* catches were predominantly fin whales, although she also caught blue, sei, humpback, sperm, and bottlenose whales.[53] In 1954, KGH decided to purchase and refurbish a Greenland Fisheries Company processing station at Tovqussaq (on the coast between Nuuk and Maniitsoq) for use as a shore-based processing plant. Tovqussaq was close to productive whaling areas, and authorities hoped that its operation would make whaling more efficient. *Sonja Kaligtoq* towed whales to Tovqussaq, where they were flensed. Workers salted the blubber for shipment to Denmark and then froze whale meat into five-kilogram packages for distribution along West Greenland's coast.[54]

KGH shut down its whaling operations in 1958 because of increasing costs and declining production. In 1959, the Tovqussaq plant was destroyed by fire. Danish authorities decided it would be cheaper to supply local communities with whale meat purchased in Norway than to rebuild their operations in Greenland. The era of Danish colonial whaling thus came to a close, but it had great impact on Greenlanders. Kapel notes that colonial whaling was influenced by three factors:

the catch per season never exceeded that necessary for consumption; the area of operation was to some extent chosen so that all districts were supplied with meat. When at last the oil production proved less profitable and the need for meat supply at the same time was considered less pronounced, the whaling operations were stopped.[55]

It was also clear that some Greenlanders resented Danish whaling. Some felt that *Sonja*'s Danish captain treated certain communities more favorably than others when making deliveries. Others resented the fact that no Greenlanders were involved in *Sonja*'s operations.[56]

In the late 1940s, these factors resulted in a new era of whaling, as Greenlanders themselves began outfitting fishing vessels with harpoon cannons. The first was the thirty-six-foot fishing vessel *Aaveq*, which

PLATE 6. Whaling aboard the vessel *Aaveq*, Disko Bay, ca. 1950
(photo courtesy of H. C. Petersen)

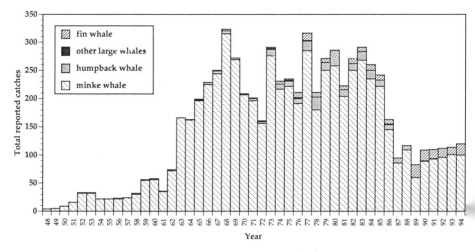

FIGURE 13. Whale catches in West Greenland, 1948–1995

began catching whales in Disko Bay in 1948.[57] Greenland's provincial council first discussed having *Sonja* catch minke whales but decided instead to outfit *Aaveq* (owned by Aasiaat fisherman Jens Geisler) with a harpoon cannon:

> . . . with reference to the hunt of small whales . . . one could not allow *Sonja* to begin hunting minke whales, but this year a harpoon cannon had been sent to the Geislers and [the speaker] awaited now the results of this experiment. It would certainly be a good undertaking for the larger motorboats, when they are first supplied with harpoon cannons, and at the same time provide the Greenlandic crews with experience in whaling.[58]

Geisler caught four minke whales in 1948 (figure 13). In 1952, he caught thirty minkes, and in subsequent years he also took the larger fin and blue whale.[59] Over time, he developed a regional market, selling whale meat and *mattak* in Aasiaat, Ilulissat, Qasiannguit, Qullissat, and Qeqertarsuaq. The vessel continued whaling on a regular basis until about 1958, when it was converted over to shrimping. During the 1950s, however, it caught an average of eighteen minkes annually.[60]

The number of Greenlandic fishing vessels equipped with harpoon cannons rose rapidly during the 1960s. From a single vessel so equipped in 1948, the number rose to forty-five by 1965.[61] The number of minkes caught from 1961–1965 also increased, from about fifty per season to about two hundred, with an average catch per vessel of about six whales. Between 1965 and the late 1970s, the number of vessels catching whales increased gradually, although not all vessels attempted whaling every year.

In about 1970, hunters in West Greenland began catching minke whales in what is known as a collective hunt. This activity is carried out using fourteen- to eighteen-foot skiffs (*umiatsiat*) equipped with outboard motors. Hunters in skiffs collaborate with others to surround a whale and shoot it with high-powered rifles. They then affix a line and floats to the whale using a hand-thrown harpoon. Collective whaling for minkes began in Disko Bay as fiberglass skiffs and powerful outboard motors became available. Outboard motors capable of keeping up with fast-swimming minkes only became available in the late 1960s. Hunters had to have cash to buy this new technology, however. Not surprisingly, collective whaling began early in the coal-mining community of Qullissat on Disko Island, where cash from wage employment was available. In 1971, hunters in Qullissat wrote a letter to the Greenland *Landsråd* in Nuuk advocating the use of this whaling technique.

This hunting method is completely new and is used not only here but also in other places. . . . [E]xperiences with minke whaling gained up to this point show that it can become a benefit for the hunting districts in the future if it is organized, as boats with outboard motors spread more and more throughout the hunting districts.[62]

The letter went on to describe a meeting of hunters in Qullissat in September of 1971 during which they agreed to self-regulate collective whaling to ensure an efficient and safe hunt. They also noted that fifteen minke whales had been caught using the technique earlier that summer.[63]

In about 1975, hunters in East Greenland also began to catch minke whales using the collective hunt. Minkes appear in the area around It-toqqortoormiit (Scoresbysund) from the middle of June until early October when sea ice begins to form. The hunt usually takes place in August and September. As in West Greenland, new outboard motors and skiffs made this technique possible. The hunt's beginnings may be tied to economic difficulties in East Greenland caused by the anti-sealing campaign of the 1970s. When hunters lost income from selling seal-skins, they were forced to look for new food sources both for their families and for dog teams.[64] Catches of minke whales in East Greenland from 1975 to 1994 are shown in figure 14.

The development of collective whaling made procurement of highly desired whale meat and *mattak* possible for the common hunter and fisherman. Households no longer had to purchase whale products from fishing vessel owners. The technique was also very compatible with the mixed subsistence-cash economies emerging in hunting districts during the 1960s and 70s. It didn't require maintenance of a large fishing vessel or a livelihood based upon full-time fishing or hunting. It could be carried out in shorter periods of time, thus meshing more easily with the demands of wage employment. It required relatively low capital inputs and was consistent with the existing hunting techniques used for beluga and narwhal hunting. The meat and *mattak* obtained in a collective hunt could be consumed locally or sold to small processing plants, like those operating in Qeqertarsuaq and Sisimiut at the time.

The collective technique revitalized small-scale whaling at the community level. As Kapel notes,

the method . . . is in accordance with the collective and co-operative way of life, which was characteristic for the hunting communities, and today needs encouragement and support. In fact, the collective catching could be regarded as a modern version of the traditional Eskimo way of hunting bowheads from umiaks [sic].[66]

The collective hunt for minkes continues today, although subject to special regulation (see chapter 4). Its contribution to the total Green-

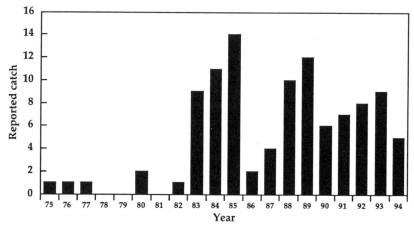

FIGURE 14. Catches of minke whales in East Greenland, 1975–1994[65]

landic catch varies from year to year, but by regulation collective hunters receive a smaller quota than do vessel hunters. In 1995, for example,
collective hunters received a quota of 52 out of the 155 whales (34 percent) allocated to West Greenland. In 1994, they received 36 percent
of the total allocation.

Contemporary fishing vessel whaling

Today, over sixty vessels in West Greenland are equipped with harpoon
cannons for taking minke and fin whales. The vessels range in size from
twenty-five to sixty feet or more and are typically geared for shrimp
trawling or other fishing. Virtually all vessels have the fifty-millimeter
Kongsberg harpoon cannon, which is mounted on the bow. Names for
the harpoon cannon and related equipment used in fishing vessel whaling are well represented in the West Greenlandic language, reflecting
the degree to which this equipment is integrated into economic life
(table 9).

The following description of fishing vessel whaling draws upon field
research in Qeqertarsuaq Municipality conducted during 1989 and
1990. At that time, only two vessels were used in whaling. Both were
used primarily for shrimping but spent one to two weeks whaling each
year. They are of two differing types: (1) an older vessel built in 1949
for use with various fishing and hunting activities; and (2) a newer vessel built in 1988, designed principally as a shrimp trawler (table 10).
These vessel types are typical of those used elsewhere in West Greenland. During 1989 and 1990, both vessels used a Kongsberg fifty-
millimeter harpoon cannon fitted with a "cold" or nonexplosive harpoon. Since 1991, all fishing vessels are required by regulation to use
harpoons fitted with the newer penthrite grenade, which kills a whale
more quickly (see chapter 4).

The older vessel in Qeqertarsuaq was owned by a father and son
who used it for catching shrimp, whales, fish, seals, and other marine
mammals. It is a North Sea–style fishing cutter, 37.6 feet in length with
19 ton displacement. Built in 1949, it was purchased by the local owners in 1984 for about $100,000. In 1992 (after these data were col-
lected), the vessel was damaged beyond repair and abandoned; nevertheless, others like it continue to be used in Greenland today. The
vessel originally had a fifty-five horsepower diesel engine, but this engine was bored out to increase the horsepower to sixty-nine. The en-

TABLE 9.
West Greenlandic and English names for whaling equipment

English name	West Greenlandic name
Harpoon cannon (complete)	*qamutilik*
Harpoon	*tuukkaq*
Cannon barrel	*savissaa*
Cannon sight (along barrel)	*nalunaqutaa*
Legs on cannon mount	*ajaappiai*
Forward part of harpoon line	*siua*
Main harpoon line	*aleq*
Float	*avataq*
Exploding harpoon	*tuukkaq qaartartulik*

TABLE 10.
Characteristics of older- and newer-type fishing vessels used in minke and fin whaling, Qeqertarsuaq Municipality, 1989–1990

Characteristic	Older-type vessel	Newer-type vessel
Year built	1949	1988
Length	37.6 feet	56 feet
Tonnage	19 BRT	46 BRT
Type of hull	Wooden	Steel
Engine type	69 hp diesel	367 hp diesel
Normal crew	4 persons	5 persons
Cost to present owner	$100,000	$875,000
Est. gross income 1989	$85,000	$550,000
Weeks fishing 1989	12	36+
Weeks whaling 1989	ca. 1	2
Harpoon type	Kongsberg 50mm	Kongsberg 50mm
Principal uses	Shrimping, whaling, seal hunting	Shrimping, whaling

gine consumes about fifteen liters of diesel per hour, and the cutter carries two thousand liters of diesel and two hundred liters of engine oil on board. The vessel has a hold capacity of fifteen cubic meters, and it is outfitted with a large shrimp trawl winch. It has both a VHF and a medium-wave radio, along with radar, a compass, and a depth finder. It can sleep four people, and the forecastle is outfitted with bunks, a table, storage cabinets, and a fisherman's stove.

The vessel's harpoon cannon was purchased used in 1984 for about $4,500 (25,000 DKK), and the cannon's mount was fabricated locally for just over $700 (4,000 DKK). The owners have six shell cases for the cannon, and they estimate that the powder, wadding, and caps used for one firing cost about $20.00. When the cold harpoon was in use, two

PLATE 7. Older-type fishing vessel used in whaling, Qeqertarsuaq

PLATE 8. Newer-type fishing vessel used in whaling, Qeqertarsuaq

harpoons were usually kept on board, both with four claws. Harpoon lines are comprised of two parts. The forerunner (*siuaa*) is seventy-five meters long and is made of thirty-five- to forty-millimeter-thick nylon line. The main line of wire cable (nine millimeters in diameter) is just over one thousand meters long. A large plastic float is attached to the harpoon line where the forerunner and the cable meet. This float serves as a brake on the whale and helps tire it out once it is hit. In 1989, the owners used the vessel for shrimping for a total of twelve weeks. Their gross income was about $63,000. In that same year, the owners spent about one week attempting to catch a minke, but their efforts were thwarted because of equipment problems.

The second type of vessel used in whaling is fifty-six feet long and is owned by five brothers from Qeqertarsuaq. The vessel was built in Denmark in 1988 for about $1 million (5.6 million DKK) to replace an older one similar to the one just described. It has a hold capacity of forty cubic meters, capable of holding ten metric tons of shrimp. The vessel is powered by a single Volvo-Penta, 367-horsepower diesel engine, which uses about thirty-three liters of diesel per hour. Tanks on board hold 8000 liters of diesel, 250 liters each of hydraulic fluid and engine oil, and one cubic meter of water. The vessel has a large deck winch for hauling shrimp trawl nets and a small deck crane. It is equipped with three radios, radar, sonar, and satellite navigation equipment. When trawling for shrimp, it normally has a crew of five or six.

Like the older-type vessel, the newer vessel's harpoon cannon is a Kongsberg fifty millimeter model. It was purchased in 1988 for about $10,900 (60,000 DKK). When whaling, the owners typically use two four-claw harpoons that cost about $1000 (6000 DKK) each. The forerunner on the harpoon line is fifty meters long and made of nylon or polyester. The main cable is fourteen-millimeter wire and is six hundred meters in length. A plastic float is fastened to the junction of the two sections.

In 1989, the vessel was involved in shrimping for about thirty-six weeks. About two weeks were spent preparing for and carrying out whaling in that year, which resulted in the catch of a single minke whale. The vessel's total gross income for 1989 was about $550,000 (over 3 million DKK). Income from the sale of minke whale meat and *mattak* comprised about 1 percent of that total, or slightly more than $5000 (30,000 DKK).

Fishing vessel whaling is highly opportunistic. Of all factors affecting success, good weather is probably the most important. Successful catches are almost always carried out in calm weather with good visibility. Fog, rough seas, or poor visibility reduce the chances of success, and vessel owners usually wait for good weather conditions. Regulations and quota limitations also influence the number of whales taken and the organization of the hunt (see chapter 4). Hunters also take economic factors into consideration. They are aware of shortages of meat and *mattak* in local communities, and they consider tradeoffs between the productivity of shrimping and whaling. Because shrimp provide the greatest income, vessel owners may hold off whaling until after the bulk of a vessel's shrimp quota has been caught. Successful whaling also depends upon having the proper equipment and a trained crew.

When a vessel owner decides to catch a whale, he typically spends little time actually searching for it; hunters generally know when and where whales can be found. The animal is usually spotted from shore or while the vessel is underway and engaged in other activities. Data gathered from hunters in Qeqertarsuaq show that nearly all spent an hour or less actually locating their catch.[67] Hunters use radios to report on the location of the whale, and other skiffs or vessels may assist in locating or tracking the whale.

Fishing vessels usually have four to six crew members on board during whaling. Typically all are men. In some cases, an experienced elder hunter may be invited along to share his knowledge. The harpooner is usually the most experienced of the crew. He directs the vessel toward the whale and must use considerable skill in firing the harpoon cannon. During the search for a whale, other crew members serve as lookouts. Sometimes crew members in skiffs accompany the vessel while whaling to assist in finding, catching, and towing the animal. Radios and walkie-talkies are used to communicate between the vessel and skiffs. After a whale is located, the time spent approaching, maneuvering, and firing the harpoon cannon can vary considerably, but in most cases it is no more than a few hours.[68] For a minke whale, the process can take only an hour or less, whereas that for the faster fin whale can sometimes take much longer.

Since 1991, vessel owners have been required by law to use the new penthrite grenade—the so-called "hot" harpoon—in whaling. The advantage of this technique is that it kills a whale more quickly than the old type. When regulations requiring the use of the penthrite grenade went into effect, the Home Rule government and KNAPK provided training courses to over 150 hunters to ensure that they knew how to use it safely and effectively.[69] According to vessel owners, one harpoon is almost always sufficient for killing a minke whale. Hunters report that the "time to death" for both minke and fin whales taken in 1994 using the hot harpoon averaged about five minutes.[70] Sometimes crew members or hunters in skiffs use rifles to finish off a whale. Once the animal is dead, its tail is bound to the vessel, and it is towed to the flensing site.

Hunters usually use well-known sites near communities to flense or butcher whales. These sites typically have a surface of smooth rock and slope gently down to the water. A whale can be towed onto this smooth surface at high tide and will be increasingly exposed as the tide drops. The rocky surface enables hunters to keep meat and *mattak* clean during flensing. In some cases, flensing also takes place on isolated beaches or on sea ice. In Qeqertarsuaq, hunters often flense whales on a rocky point just outside of the community's harbor. There, hunters use a hand winch to haul whales up onto smooth rock at the water's edge. In Kangerluk, whales are often flensed at a beach site not far from the settlement. Flensers typically use large kitchen knives to cut up the slabs of meat and *mattak*. The process usually takes about three to four hours for a minke whale, depending upon the number of participants and the weather conditions. Flensing the larger fin whale may take from six to ten hours or more.

The social organization of fishing vessel whaling is usually centered around the nuclear or extended family. Most vessels used in whaling are owned by family units, even if it comprises several households. This pattern differs from that involved in large-scale offshore shrimping, where kinship is less significant. Figure 15 illustrates kinship relationships among whaling crew members in 1988 on the older-type vessel described above. The owners (father and son) were joined by the former's son-in-law and by a young son of another daughter. A total of eight people participated in the flensing, most of whom were related to the elder hunter.

Collective whaling for minke whales

In a collective hunt, participants in small skiffs with outboard motors surround a minke whale and shoot it with rifles. When the whale slows, it is harpooned and finally killed with rifle shots. Because fin whales are so much larger, regulations prohibit hunters from using the collective

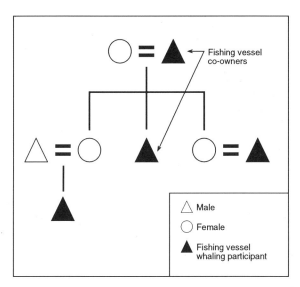

FIGURE 15. Social organization of fishing vessel whaling, Qeqertarsuaq, 1988

technique to catch them. Typically, fiberglass skiffs are used in collective whaling, ranging in size from fourteen to eighteen feet. Hunters in Qeqertarsuaq commonly use a Mariner or Yamaha forty-horsepower outboard motor. Larger outboards are occasionally used. Table 11 shows the characteristics of equipment used in collective hunts in Qeqertarsuaq, including the average numbers of skiffs and participants involved. The number of skiffs participating averages sixteen, with a range from five to thirty-five. In recent hunts, an average of thirty hunters have participated, generally with two men per skiff. Most hunters use a .30–06 rifle, and nearly all skiffs carry a harpoon with line and several large plastic floats attached. The harpoon head is commonly homemade, with a metal point and wooden shaft. Hunters use both walkie-talkie radios and hand signals to communicate about the whale's location and to coordinate their efforts.

Importantly, most households already own the equipment used in collective whaling. This fact enables far more hunters than just those who own vessels to participate in a hunt. Furthermore, the equipment used costs considerably less than that employed in vessel whaling (table 11). In a single hunt, participants spend an average of about $53.00

TABLE II.
Characteristics of equipment and participants in collective minke whaling,
Qeqertarsuaq Municipality, 1989

Characteristic	Description
Most common length of skiff	14 foot
Average number of skiffs participating	16
Average number of hunters participating	30
Average number of hunters per skiff	2
Range of skiffs participating	5 to 35
Most common rifle caliber	.30–06
Average expenses per hunter for fuel and ammunition	$52.91 (293 DKK)

PLATE 9. Participants in collective hunt for minke whale

(292.52 DKK) for outboard gas (an estimated forty liters) and oil and for ammunition.

As in vessel whaling, the collective hunt is largely opportunistic. Hunters are generally aware when whales are in the area, and the presence of whales near a settlement is a common topic of conversation. In Qeqertarsuaq, many families have a good view of the ocean from their home and frequently scan nearby waters with binoculars. In other cases, hunters may encounter a minke whale while out fishing or hunting.

Calm winds and good visibility are essential for a successful hunt. The speed and indistinct blow of a minke whale makes tracking the animal difficult even under the best conditions. Shooting accurately from a skiff is always difficult. Hunter safety is a serious concern in the collective hunt. In the winter of 1989–90, a bullet fired in one such hunt for beluga ricocheted off the water and hit another hunter, wounding him seriously. In hunts having as many as sixty participants, great care must be taken.

Once a minke whale is spotted, hunters usually communicate by radio about its location. If enough skiffs and qualified hunters are available to participate, the whale is pursued (collective hunters must be licensed in advance to participate). As in vessel whaling, little time is actually spent searching for a whale. In most cases, it has already been spotted, and participants join in after getting the word about its location. Hunters kill a whale by maneuvering skiffs into position alongside it when it surfaces and by shooting it with rifles, aiming for the lungs. As the whale dives, hunters keep close watch, trying to anticipate where it will surface next. Once the whale has been slowed by bullets, hunters attach a harpoon, line, and floats to tire it. This action also helps ensure that when it is dead, the whale will not sink.

The process of shooting and harpooning a minke whale usually takes one hour or less. Hunters in 1994 reported that the average "time-to-death" in the collective hunt was thirty minutes.[71] Once the whale is dead, a line is attached to its tail and several skiffs work together to tow it to the flensing site. Towing the whale with skiffs can be a slow process, depending on the number of boats involved, the distance to shore, and sea conditions. Hunters use essentially the same technique for flensing a minke whale caught in a collective hunt as they do in fishing vessel whaling. The major difference in the collective hunt is that typically there are far more participants; the process thus goes more quickly, taking three hours or less.

Kinship is a major factor determining who participates in collective whaling, just as it is in vessel whaling. Figure 16 illustrates kinship relationships in one such hunt in Kangerluk in 1988. Participants came from the two major extended families in the community. Experienced hunters often play an important role in a successful hunt. Younger hunters learn by observing elders, deferring to their knowledge and experience in making decisions. The prestige of participating in whaling and the satisfaction of obtaining food for one's family are strong inducements for younger hunters to take part.

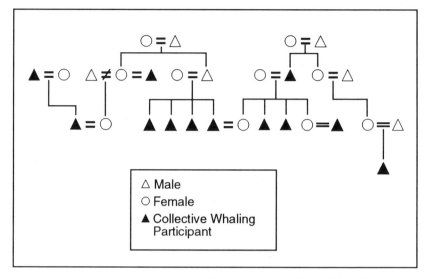

FIGURE 16. Social organization of collective whaling, Kangerluk, 1988

Household participation in minke and fin whaling is extensive in many Greenlandic communities. In Qeqertarsuaq Municipality, for example, nearly 70 percent of all surveyed households said members had done so recently.[72] Most had participated in collective rather than vessel hunts, reflecting the fact that relatively few own vessels. Hunters participated in collective whaling with remarkable frequency; on average, about once a year.[73] In contrast, households participated in fin whaling much less frequently.

Consumption, distribution, and exchange of whale products

Greenlandic beliefs and traditions shape patterns of consumption, distribution, and exchange of whale products. As with other country foods, whale meat and *mattak* are widely consumed in local communities. For example, in Qeqertarsuaq and Kangerluk, fully 97 percent of all households use minke whale products, and 73 percent use those from fin whales.[74] Minke and fin whale meat is cooked in stews, fried in butter, or eaten dried (*nikkut* in West Greenlandic). *Mattak* is generally eaten raw after having been frozen, or it is boiled. Fin whale oil is sometimes eaten with dried meat or fish. Whale meat is also used by some households for sled dog food, particularly in northern West Greenland and in East Greenland. About one-quarter of all households surveyed in Qeqertarsuaq Municipality use it for dogs.

Once whales are flensed, products from them enter into a complex distribution network where economic considerations, social relationships, and cultural norms all interact. Cash is an element in this network, but its significance is limited. It is but one part of the household strategies hunters use to procure foods and other goods that keep social relationships viable.[75] After a whale is flensed, the products from it are distributed in at least three stages: (1) hunters themselves divide products among those directly involved in the catch; (2) hunters' households share products with other households, or may sell them for cash; and (3) recipient households may in turn share products with others. In vessel whaling, the owner(s), crew, and flensers share in the first stage of distribution. In addition, the vessel itself receives a share, generally 40 to 50 percent, to pay for fuel and other expenses.

Figure 17 shows how hunters in Qeqertarsuaq distributed some two thousand kilograms of meat and *mattak* taken from a minke whale in 1988. After the whale was flensed, meat and *mattak* were piled into fish boxes, each of which holds about 70 kilograms. The vessel's owner and the captain (who are brothers) each received two boxes (140 kg each). Each of the additional three crew members received one box (70 kg); furthermore, three other brothers of the owner and captain (all co-owners of the vessel) each received one box (70 kg). In all, about 700 kilograms of meat and *mattak* were distributed among the owner's family of five brothers and among the crew. The remaining 1300 kilograms of meat and *mattak* were then sold at the *kalaaliaraq* to pay for expenses. The exact proportions of meat and *mattak* sold are not known, but if we assume that 1200 kilograms were meat and 100 kilograms were *mattak*, the total return would be about $5700 (about 35,000 DKK). In this case, the circle of direct recipients of whale products is relatively small; however, at the second and third stages of distribution, households distributed shares widely to other kin and acquaintances in the community. Households without direct access to shares were able to purchase products at the *kalaaliaraq*.

The distribution of minke products from collective whaling involves a special method of dividing up the catch. It typically includes a much wider circle of recipients in the community. As mentioned, as many as

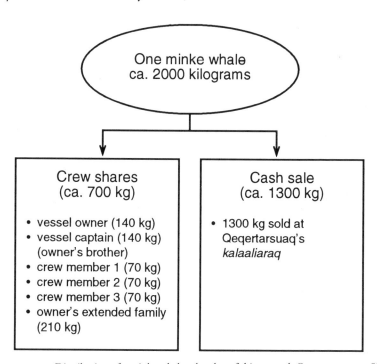

FIGURE 17. Distribution of a minke whale taken by a fishing vessel, Qeqertarsuaq, 1988

thirty or more skiffs and fifty to sixty hunters might participate in a collective hunt. Each skiff is entitled to an equal share of the whale. A special "blind" process is used by hunt participants to ensure that no one share is larger than the other. In this process, equal piles of meat, *mattak*, and *qiporaq* are created at the flensing site, one for each of the skiffs actively participating in the hunt (figure 18). Once the flensing is completed and all hunters are satisfied that the piles are equal, one hunter from each skiff stands in a line facing two men who serve as distributors of the whale. The first of the distributors (the *taaguisoq*, or "one who names") stands so that he sees only the line of hunters and not the piles. The other distributor (the *tikkuartuisoq*, or "one who points") stands behind the first and points to a particular pile at random. As he points, the first distributor calls out the name of one hunter in line, who then collects the pile of whale products for himself and the partner(s) in his skiff. Using this system, all are assured of equal treatment because the man calling out the name of the hunter has no idea which pile is being pointed out. After each skiff receives its share, the hunters involved typically divide the products equally between themselves. In some cases, the person who paid for gas, oil, and shells receives additional compensation.

In the second stage of distribution, collective whaling participants typically share with other family members and may sell some or all of their share at the *kalaaliaraq*. In the example shown in figure 18, twenty-four skiffs participated in the hunt. Each received about eighty kilograms of whale produce. In the example shown in the figure, a hunter used forty-five kilograms of his share for his household and sold the rest at the *kalaaliaraq*. At the third stage of distribution, households receiving shares from hunters often distribute again to others. For example, one household in Qeqertarsuaq sent *mattak* to a member of the family attending school in Nuuk.

As described in chapter 2, hunters can sell whale products for cash at the local processing plant, at the *kalaaliaraq*, or through private sales (table 12). Prices for these products are fixed, generally on an annual basis, through negotiations between Royal Greenland and KNAPK (the national hunters' and fishers' association). Prices charged locally

PLATE 10. "Blind" system used by hunters to distribute whale products

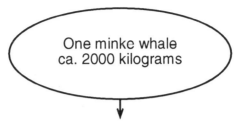

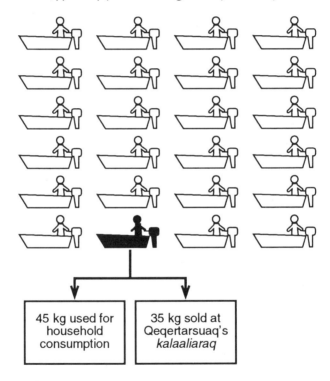

One minke whale
ca. 2000 kilograms

24 skiffs participating in collective hunt

*Each skiff receives equal shares of meat, mattak, and
qiporaq (ca. 80 kilograms per skiff)*

45 kg used for
household
consumption

35 kg sold at
Qeqertarsuaq's
kalaaliaraq

FIGURE 18. Distribution of minke whale products from a collective hunt, 1988

at the *kalaaliaraq* are set by agreement between the municipality and the local hunters' organization.

Hunting success, regulatory constraints, and the availability of alternative resources (including cash) all influence whether a household sells whale products. Table 13 shows the amounts of minke and fin whale products sold to Royal Greenland in the period from 1987 to 1993.[76] In general, these sales are rather sporadic. For example, in the years from 1987 to 1992, whale products were sold to Royal Greenland in only one municipality (Qaqortoq). In other years, hunters sold products in Nuuk and in Maniitsoq. The same is true for sales at the *kalaaliaraq*. In 1989 and 1990, only 2 percent of households surveyed in Qeqertarsuaq sold whale products (in this case totaling about $1270 or 7000 DKK). There are several reasons why sales are so limited. In

TABLE 12.
Comparative prices for minke whale products (in US$) through Royal Greenland and at Qeqertarsuaq's kalaaliaraq, 1990

Product	Royal Greenland price ($ per kg)	*Kalaaliaraq* price ($ per kg)
Minke whale meat (fresh)	1.81	4.00
Minke whale *qiporaa* (fluted belly flesh)	2.72	5.46
Minke whale *mattak* with blubber	1.14	2.73
Minke whale *mattak* without blubber	Not available	9.09
Dried minke whale meat (*nikkut*)	7.82	Not available

PLATE 11. Selling minke whale meat at the *kalaaliaraq*

TABLE 13.
Total sales of minke and fin whale products to Royal Greenland A/S, 1987–1993

	1987	1988	1989	1990	1991	1992	1993
Kilos sold	3,226	15	12,428	38,916	22,001	37,216	103,791
Value (DKK)	24,201	266	124,779	407,968	224,536	384,389	1,204,630
Ave. pr. kg	7.5	17.73	10.04	10.48	10.20	10.32	11.60

SOURCE: Greenland Home Rule Government 1994.

many cases, there are not enough whale products to meet local needs. Households surveyed in Qeqertarsuaq in 1990, for example, cited low IWC quotas as the major reason that more products were not available.[77] In some cases, households equipped to participate in collective hunting are unable to do so because of quota or licensing limitations at the municipal level. In other cases, collective hunters find that their minke quota is used up even though a quota allocation is still available for vessel owners to take a fin whale.

Whaling and Greenlandic identity

Greenlandic society today is undergoing substantial change through interaction with the world economy. Market prices for everything from shrimp to sealskins to heating oil influence the lives of Greenlanders, even in the most remote settlements. Although ripples of change course through community life, certain activities provide a sense of continuity and an affirmation of what it is to be a Greenlander in a turbulent world. Hunting marine mammals and sharing the food from them is one of these. Participation in hunting provides Greenlanders with a connection to a way of life reaching back over four thousand years, a connection shared with fellow Inuit in Canada, Alaska, and Chukotka. As an earlier IWC report notes, the Inuit as a people are "by tradition the most hunting-oriented of all human groups, because their environment provides very few non-animal resources."[78] Whaling and the sharing of country foods are thus manifestations of cultural continuity with Inuit traditions of great depth.[79]

As Dahl points out, whaling and marine mammal hunting in Greenland serves complex integrative and cultural functions.[80] In Qeqertarsuaq, the social organization of whaling remains closely tied to kinship. When extended family members participate in whaling, whether on a fishing vessel or in a collective hunt, they strengthen the bonds of kinship through shared experience. This participation initiates a host of social and cultural interactions. As family members assist in flensing, they share meat and *mattak* and they regale each other with stories about the hunt. The language of these interactions is *Kalaallisut*. Much has been written about relationships between language and culture in Greenland, focusing on the central role of shared language in affirming a collective identity. When hunters talk excitedly amongst themselves over walkie-talkies about sighting a whale, they share that discovery with a discrete cultural group because of their common language. Whaling reinforces these connections by employing words and phrases about the hunt that were developed far back in Inuit history.

Similarly, whaling is based on shared knowledge, values, and beliefs. When an elder is along on a hunt, his knowledge about whale behavior and characteristics is respected. Elders transmit culturally appropriate behavior to others through example. They admonish younger hunters to use the animal thoroughly and to avoid waste. In this manner, Inuit knowledge and belief systems are passed on to younger hunters through example and experience.

Whaling is also a source of prestige. Greenlandic men speak with a barely muted pride when they describe hurling a harpoon at a whale. This pride is distinctly male, because women rarely participate. Through the family's celebration of a boy's first seal to the excitement of a community upon hearing that a whale has been taken, the male's role as hunter continues to be validated. This validation may be especially important in communities like Qeqertarsuaq, where many men are engaged in wage employment. Participation in a collective hunt provides a touchstone for validating one's capabilities as a hunter, even as one spends much of the day in an office or a shrimp processing plant.

At the community level, whaling contributes to mutual security for all. Men who are seasonally unemployed, or who can't find work at all, know that they and their family can survive because they can hunt seals and whales both for food and for cash. In Qeqertarsuaq, the community's esteem for whales and whaling is reflected in the use of the bowhead whale as the municipal crest or shield. Most historic structures in the town are associated with whaling, including the beehive-shaped lookout on *Qaqqaliaq*, a small nearby promontory overlooking the sea. The structure, built originally with four bowhead whale jawbones, served as a protected lookout for whalers. In the mid-twentieth century it blew down in a powerful storm, and efforts were made to rebuild it in time for the municipality's two-hundred-year jubilee in 1973. In 1968, local authorities received special dispensation to catch two bowhead whales so that this historic structure could be replaced. Only one whale was caught, however, and the lookout had to be reconstructed with wood. The jawbones from the one whale now serve as a ceremonial archway in Qeqertarsuaq's harbor.

Whaling also contributes to an emerging Greenlandic national identity. As international conflicts over marine mammal hunting continue, Greenlanders feel a growing sense of solidarity in fighting to protect what they view as a fundamental right to use local resources sustainably. These conflicts also contribute to pan-Inuit solidarity, with Greenlanders joining fellow Inuit at IWC meetings to emphasize the importance of whaling to Inuit culture. It is to these conflict-filled relationships between hunters, the Home Rule state, the IWC, and the global community that I now turn in chapter 4.

≈≈ 4 ≈≈

Greenland's Whaling Regime: Costs and Benefits of Co-management

C o-management, as defined in the introduction, means sharing power and authority between resource users and the state, often at multiple levels. A co-management *regime* involves agreed-upon principles, norms, rules, and decisionmaking procedures for managing renewable resources. Greenland's whaling regime incorporates a wide array of stakeholders, institutions, and processes at the local, national, and international levels. This chapter describes the historical development of this regime, the stakeholders involved, and the conflicts surrounding whaling today. It focuses particularly on contradictions that Greenlanders face in using whales sustainably, including those having to do with conflicts over quota allocations, use of new hunting technologies, validity of scientific data, and the morality of exchanging whale products for cash.

Stated simply, co-management of whaling in Greenland means that the IWC establishes catch quotas for minke and fin whales and leaves day-to-day management to the Danish and Home Rule governments. Table 14 shows the history of quota allocations from the IWC to Greenland for the period 1984 to 1995. In 1995, West Greenlanders were allowed to catch 155 minke whales from a three-year quota of 465. They were also allowed 19 fin whales. East Greenlanders were allowed to catch 12 minke whales.

Although on the surface Greenland's relationship to the IWC may seem straightforward, in reality it is far more complex. Indeed, this relationship reflects a long history of political marginalization, wherein indigenous whaling practices were undermined as other nations overexploited whale stocks. In the past, Greenlanders had little impact on the viability of these stocks because hunters were small in number and had limited technology. As we've see, this situation began to change as

TABLE 14.
IWC quotas for Greenlandic aboriginal subsistence whaling, 1984–1995

Year	Minke Whales		Fin whales	Humpback whales
	W. Greenland	E. Greenland		
1984	2-yr. quota total = 588 max. 444 per year	10	6	9
1985		10	8	8
1986	2-yr. quota total = 220 max. 130 per year	10	10	0
1987		10	10	0
1988	110	12	10	0
1989	60	12	23	0
1990	2-yr. quota total = 190 max. 100 per year	12	2-yr. quota total = 42 max. 23 per year	0
1991		12		0
1992	3-yr. quota total = 315 max. 105 per year	12	21	0
1993		12	2-yr. quota total = 42	0
1994		12		0
1995	3-yr. quota 1995–97 = 465 caught/struck max. 155 per yr.	12	19	0

Greenland was incorporated into the Danish realm. Today, Greenland participates in the IWC as a part of Denmark. Under Home Rule, hunters must follow Greenlandic law and regulations that implement quotas. As we shall see, these requirements have both advantages and disadvantages. The tensions created by working within this regime illustrate both the strengths and the limitations of Home Rule and the costs and benefits of co-management.

Aboriginal subsistence whaling and the IWC: A brief history

When Greenlanders today participate in the IWC, they do so in the belief that indigenous rights to whaling are protected under international law.[1] In Euro-American law, articulation of these rights dates back to three legal scholars from the sixteenth and seventeenth centuries: de Vitoria (1492–1546), Grotius (1583–1645), and Pufendorf (1632–1694).[2] Their work, and subsequent developments in international law, provide the basis for Article I, Section 2 of the International Covenant on Economic, Social and Cultural Rights, which reads:

All people may, for their own ends, freely dispose of their natural wealth and resources without prejudice to any obligations arising out of international economic co-operation, based upon the principle of mutual benefit, and international law. In no case may a people be deprived of its own means of subsistence.[3]

The covenant also refers to the "right of everyone to the enjoyment of the highest attainable standard of physical and mental health" and to every person's right "to take part in cultural life." Similarly, the International Covenant on Civil and Political Rights and the International Labour Organization (ILO) Convention no. 169 affirm that minorities shall not be denied the right to enjoy their culture. Greenlanders point to these principles as the basis for their rights to use marine mammals sustainably.[4]

The International Convention for the Regulation of Whaling (ICRW), signed by fifteen governments in December of 1946, provides the basis for the current aboriginal subsistence whaling regime. The convention established the IWC to "provide for the proper conservation of whale stocks and thus make possible the orderly development of the whaling industry." Membership in the IWC is open to any country adhering to ICRW principles; it currently has thirty-six members. The IWC has three committees that provide recommendations within their area of expertise: Scientific, Technical, and Finance and Administration. The Technical Committee has two standing subcommittees, one for aboriginal subsistence whaling and one for infractions (violations of regulations). The Scientific Committee provides advice on the biological status of whale stocks, which forms the basis for Technical Committee recommendations to the IWC plenum about the management of whaling. Changes to the ICRW Schedule (which contains quotas for whaling) require a three-quarters majority vote.

The reader interested in a more comprehensive account of the IWC's history and of conflicts associated with it can find ample published material elsewhere.[5] The material presented here is limited to the aboriginal subsistence whaling regime itself. The ICRW was signed primarily to address issues of commercial whaling;[6] however, a key provision states: "It is forbidden to take or kill gray whales or right whales, except when the meat and products of such whales are to be used exclusively for local consumption by the aborigines." This exception for "aborigines" reflects provisions of an earlier convention signed in 1931. This earlier agreement states:

The present Convention does not apply to aborigines dwelling on the coasts of the territories of the High Contracting Parties provided that:
1) They use canoes, pirogues or other exclusively native craft propelled by oars or sails.
2) They do not carry firearms.
3) They are not in the employment of persons other than aborigines.
4) They are not under contract to deliver the products of their whaling to any third person.

This exemption was clearly designed to limit external influences on indigenous whaling by controlling both technologies and interaction with distant markets.

In the years following adoption of the 1946 ICRW, additional provisions were added to the Schedule to accommodate aboriginal subsistence whaling.[7] For example, in 1961 member nations enacted a special provision accommodating pre-existing humpback whale catches by Greenlanders despite the fact that commercial catches were prohibited. In 1964, the IWC amended the Schedule to ensure that gray and right whales could only be taken for local consumption by aboriginal peoples, or on their behalf.

In 1972, however, the UN Conference on the Human Environment popularized the idea that ocean resources are a "global commons"; that is, the responsibility of all nations. This concept, propagated during the height of the environmental movement in North America and Europe, reflected new discourses about the use and preservation of marine mammals and other ocean resources. Although it did not address aboriginal subsistence whaling, the conference approved a resolution calling for a ten-year moratorium on commercial whaling. A decade later, the 1982 UN Convention on the Law of the Sea (UNCLOS) accepted the idea that certain ocean resources are the "common heritage of mankind."[8] Proponents argued that "it should be understood as a wholly new concept of property rights, a modern alternative to the traditional ideas of exclusive ownership or of free and unlimited access."[9] Significantly, however, UNCLOS negotiators chose to limit application of the "common heritage" concept to seabed minerals and not to apply it to whales or other marine mammals. As De Klemm points out, this restraint occurred because "states are now afraid that in an open convention, a majority of parties that are not exploiting a resource would be able to stop all harvesting against the will of the minority."[10]

In 1975, the IWC strengthened its conservation policies by adopting the so-called "New Management Procedure." It adopted more precise regulations for whaling on stocks considered to be within the "protection stock" category and for achieving optimum catch levels for abundant stocks.[11] The IWC also recognized the need for a more specific management regime for aboriginal subsistence whaling. This need became apparent when, in 1977, the Scientific Committee expressed concern about Alaskan Eskimo bowhead catches under the aboriginal subsistence whaling exemption. Alaskan Eskimos have strong cultural and nutritional ties to bowhead whales, but stocks were severely depleted by American whalers in the latter part of the nineteenth century. During the 1970s, the total number of bowheads caught by Eskimos, and particularly the number struck but lost, grew significantly. Reasons for this increase in hunting are complex, but it is generally attributed to cultural revitalization efforts, declining availability of other country foods (especially caribou), changing hunting technologies, and growing access to cash through oil development and the Alaska Native land claims settlement.[12]

Responding to the Scientific Committee's concern about bowheads in Alaska, the IWC deleted the aboriginal subsistence exemption for all right whales (including bowheads) at its 1977 annual meeting. This measure unleashed a storm of controversy, with Alaskan Eskimo whalers threatening to defy international regulations to protect subsistence rights. Hunters reacted swiftly to form the Alaska Eskimo Whaling Commission (AEWC), comprised of whaling captains from nine (later ten) communities. The AEWC argued that the IWC and the US government had acted precipitously and that biologists knew little about the bowhead stock and had underestimated its size. The crisis was resolved only after the US government agreed to push for a modest take of bowheads (despite significant pressure from whale preservation interests), coupled with an intensive management scheme involving research and monitoring. Alaska's North Slope Borough also decided to invest considerable funds in bowhead research.

The IWC response to the Alaskan bowhead crisis was to form a special working group of the Technical Committee in 1978. The working group was to examine the entire aboriginal whaling problem and to develop proposals for a regime applicable to the Alaskan situation and, if appropriate, for other aboriginal hunts.[13] The working group convened three panels of experts in 1979 and produced an IWC report on aboriginal subsistence whaling in 1982.[14] By 1980, it was clear that the

IWC needed a more comprehensive evaluation of management principles and guidelines for subsistence catches. It therefore established a working group of the Technical Committee that brought together not only IWC delegates (including some indigenous people) but also representatives of indigenous organizations. The working group met in 1981 and adopted three important definitions, which the IWC as a whole later adopted:

Aboriginal subsistence whaling means whaling for purposes of local aboriginal consumption carried out by or on behalf of aboriginal, indigenous or native peoples who share strong community, familial, social and cultural ties related to a continuing traditional dependence on whaling and on the use of whales.

Local consumption means the traditional uses of whale products by local aboriginal, indigenous or native communities in meeting their nutritional, subsistence and cultural requirements. The term includes trade in items which are by-products of subsistence catches.

Subsistence catches are catches of whales by aboriginal subsistence whaling operations.

The group also agreed that:

[T]he full participation and involvement of the indigenous peoples are essential for effective whale management; and that it is in the best interests of all three parties involved (the IWC, the national governments and the indigenous people) to involve the indigenous people in the decision-making process.[15]

Controversy developed within the group over management principles, with some members arguing against a distinction between aboriginal and commercial whaling. Most accepted that commercial whaling focuses on maximizing yields from individual stocks, whereas subsistence whaling focuses more on obtaining sufficient whale products to satisfy nutritional and cultural needs. Despite disagreements, the group agreed on three broad objectives in managing aboriginal subsistence whaling:

To ensure that the risks of extinction to individual stocks are not seriously increased by subsistence whaling;

To enable aboriginal people to harvest whales in perpetuity at levels appropriate to their cultural and nutritional requirements, subject to the other objectives; and

To maintain the status of the whale stocks at or above the level giving the highest net recruitment and to ensure that stocks below that level are moved towards it, so far as the environment permits.

The group also (1) recommended a procedure for establishing catch limits in subsistence hunts; (2) agreed that commercial whale catches should not interfere with subsistence needs; (3) endorsed the need for enhanced research, reporting, and monitoring; and (4) supported efforts to make subsistence hunts as efficient and humane as possible, consistent with cultural traditions and practices.[16]

As a result of the working group's efforts, the IWC adopted a resolution in 1982 implementing the current aboriginal subsistence whaling management regime (also referred to as the "Aboriginal Whaling Scheme"). It amended the ICRW Schedule to specify the following criteria for setting subsistence catch limits:

(a) Notwithstanding the provisions of paragraph 10 [which sets out the management principles for commercial whaling], catch limits for aboriginal subsistence need for the 1984 whaling season and each whaling season thereafter shall be established in accordance with the following principles:

(1) For stocks at or above the Maximum Sustainable Yield (MSY) level, aboriginal subsistence catches shall be permitted so long as total removals do not exceed 90 per cent of MSY.

(2) For stocks below the MSY level but above a certain minimum level, aboriginal subsistence catches shall be permitted so long as they are set at levels which will allow whale stocks to move to the MSY level.*

(3) The above provisions will be kept under review, based upon the best scientific advice, and by 1990 at the latest the Commission will undertake a comprehensive assessment of the effects of these provisions on whale stocks, and consider modifications.

*The Commission, on advice of the Scientific Committee, shall establish as far as possible (a) a minimum stock level for each stock below which whales shall not be taken, and (b) a rate of increase towards the MSY level for each stock. The Scientific Committee shall advise on a minimum stock level and on a range of rates of increase towards the MSY level under different catch regimes.[17]

The commission established a standing Sub-committee on Aboriginal Subsistence Whaling under the Technical Committee to consider documentation on nutritional, subsistence, and cultural needs and to provide advice about management. In adopting the resolution, the IWC also specified which stocks are covered by the subsistence regime. These have been amended since 1982 and now include:

1. Bering-Chukchi-Beaufort Seas stock of bowhead whales taken by Alaskan Eskimos

2. minke and fin whales from the West Greenland stocks of these species, and minke whales from the central North Atlantic stock (taken by Greenlanders)
3. eastern North Pacific gray whales (taken by the native peoples of the Russian Far East and by Alaska natives)
4. humpback whales by the Bequians of St. Vincent and The Grenadines in the Caribbean

In 1982, the IWC decided to impose a temporary pause or moratorium on commercial whaling. Responding to uncertainties about the status of some whale stocks and to pressure from anti-whaling forces, it determined that all catch quotas would be set at zero after the 1985/86 season. At the same time, the IWC agreed to conduct a comprehensive assessment of the effects of the zero quotas by 1990, when modification would be considered.

Although several member nations initially used their right to object to the zero quotas, all commercial whaling of species covered by the IWC was suspended during this pause. Some whaling continued for a time under provisions for scientific whaling. In 1993, the Scientific Committee recommended unanimously that the IWC adopt the so-called "Revised Management Procedure" (RMP) and its associated "catch limit algorithm" (CLA), a mechanism for calculating quotas for commercial whaling. The RMP is designed in principle to provide a balance between conservation and utilization of baleen whales and to determine sustainable catch limits given limited data and uncertainties about ecological dynamics. Its goal is to manage whale stocks at approximately 72 percent of their carrying capacity or pre-exploitation size. It prohibits catches from stocks below 54 percent of the pre-exploitation level and provides mechanisms for regularly updating data about the stocks.[18] In 1994, the commission adopted the procedure after several additional tests were completed;[19] however, new whaling quotas have not yet been adopted based on the RMP and CLA.[20]

In the minds of some, approval of the RMP sets the stage for a resumption of commercial whaling on a sustainable basis. Others believe, however, that the adoption of the RMP is simply one stage in a widening debate about the future of whaling. They point to growing opposition to whaling on ideological or ethical grounds.[21] No matter what the outcome of this contentious debate, the IWC's approval of the RMP for commercial whaling may also lead to revision of procedures for aboriginal subsistence whaling. Ray Gambell, the IWC's secretary, notes that once the IWC completes its current review of procedures for commercial whaling, it may well turn its attention to aboriginal subsistence whaling:

[The Scientific Committee] assumed that any revised procedure for commercial whaling would be generally compatible with that for aboriginal subsistence whaling and that a full discussion of any new scheme for the latter could only usefully take place after an alternative management procedure for commercial whaling had been established.[22]

Indeed, in 1994 the IWC adopted a "Resolution on a Review of Aboriginal Subsistence Management Procedures," calling for the Scientific Committee to "investigate potential management regimes for aboriginal subsistence whaling, including regimes based on the approach taken in the Revised Management Procedure and utilizing simulation trials where appropriate."[23] The resolution affirms that the committee's review should be based on the three principles for subsistence whaling adopted in 1982; furthermore, the review is to give highest priority to the objective of "ensuring that the risks of extinction to individual stocks are not seriously increased by subsistence whaling." In 1995, the committee responded that this issue was actively being discussed but that further development of an alternate Aboriginal Whaling Scheme would likely require creating a separate steering group and convening a series of workshops on the topic.[24]

Although it remains unclear what relationship there might be between the RMP and any changes to the aboriginal scheme, Gambell believes that the procedure now in place for aboriginal whaling is insufficient:

The current regulations unfortunately request scientific input that is not within our grasp to provide, and the procedure presently followed in practice places the primary determination of catch levels on the perceived subsistence need of the local human populations, with rather little weight attached to the biological capacity of the whale stock to sustain that amount of catch. . . . It would at all events be good for a more practical procedure to be developed that takes account of all the relevant factors, both with respect to the status of the stocks of the whales themselves and to the subsistence needs of the aboriginal hunters and their communities.[25]

Indigenous leaders, however, express deep concern about using the RMP as the basis for management, believing it could lead to sharp quota reductions.[26] Ingmar Egede, until recently Greenland's vice-president of the Inuit Circumpolar Conference (ICC), notes that some nations opposed to commercial whaling are now advocating that the Revised Management Procedure (RMP) or some variant of that exceedingly conservative quota setting approach be applied to aboriginal whaling. If current experiences with the application of the RMP are any indication, management decisions would be further removed from any direct influence by community based whalers to the serious detriment of their traditional knowledge, cultural traditions, and socio-economic needs.[27]

These concerns arise not simply from formal actions taken within the IWC but also from indigenous peoples' years of experience with the IWC's intensely political environment. Caleb Pungowiyi, formerly president of the ICC, describes his impressions from the 1994 IWC meeting:

We Inuit have gradually become the only ones on this earth who have a hunt for large whales. Because of this, we're being placed under a microscope by the world community and watched very carefully.... Any little mistake we make, any little evidence of waste of the resource, any "inhumane" treatment of the whales. Everything will be picked up by the rest of the world.... For some people, our whaling is repulsive and is seen as a direct threat to the whales' survival.[28]

Nancy Doubleday goes further in accusing the IWC of "collective amnesia" by ignoring the efforts of the working group in 1981 and 1982. Her experiences of working with Inuit people in the IWC context lead her to believe that

in actual practice, a coalition of anti-whaling nations operating in a clandestine fashion has circumvented the letter and the spirit of the Working Group recommendation that full participation and involvement of the indigenous peoples are essential for effective whale management.... Calling itself the "Like-minded Group", this coalition of anti-whaling nations which hold the majority in the IWC meet in private to make the deals that ultimately become the "decisions" of the IWC in subsequent sessions of the Commission. Indigenous peoples are not welcome at these "Like-minded Sessions". In this way, the advances made by indigenous peoples in the IWC itself are undermined by the anti-whaling interests.[29]

These concerns are compounded by charges that the IWC's aboriginal subsistence whaling category is at best inappropriate and at worst racist in nature. Several nations initially opposed establishing such a category at all. More recently the issue has come up in debates about "small-type coastal whaling," a proposed category that could perhaps apply to community-based whaling in Norway, Iceland, and Japan. Kalland, for example, sees striking similarities between aboriginal subsistence whaling and whaling by nonaboriginal peoples in other coastal communities.[30] He objects to the aboriginal whaling concept because he believes it is based on false premises:

> . . . concepts such as [aboriginal subsistence whaling] imply a static view of a people and its culture. Whaling—as well as sealing—is allowed only as long as it is conducted by small non-white, oppressed minorities perceived as lacking unifying political institutions, using "simple" technologies, and whose economic exchanges are believed to exist within the confinement of a non-commercial economy. Only "traditional" usage is allowed, and it tends to be the outsiders who define what is "traditional." To allow whaling and sealing under the above conditions gives the anti-whalers a way to control ethnic minorities and keep them in a position of dependency.

Kalland argues that aboriginal subsistence whaling is a "concept in the service of imperialism" because it is used to limit whaling practices and to divide indigenous whalers politically from others who wish to undertake whaling on a sustainable basis. In his view, permission to catch whales should be determined by whether or not the activity is sustainable rather than on the basis of ethnic or cultural origins. Some in Greenland acknowledge the ethnic basis of the IWC's current whaling regime but see little alternative given the politically charged environment in which decisions are made. Speaking anonymously, one Greenlander involved in whaling issues commented:

> In my view, ethnically-framed categories are completely contrary to the principle of sustainable use of living resources, and are in fact fundamentally wrong. But what would it mean if Greenland were lumped under a small-type whaling category? There would be problems with quotas. We really have no alternative but to protect our current status.

I will return to these contentious issues in later chapters after examining the historical development of Greenlandic whaling policies in this century.

Marginalization of Greenlandic whaling, 1931 to 1979

Greenland's history of marginalization within the IWC originated in the era of Danish colonialism. Indeed, when Denmark signed the ICRW in 1950, it seemed unaware that the convention would affect Greenland. Jens Brøsted, a scholar in Danish-Greenlandic legal affairs, documents how mistakes and administrative arrogance by colonial authorities steadily marginalized Greenlanders' political position.[31] According to his account, Greenland's two *landsråd*, or provincial councils, established at the turn of the century, were to review all matters affecting the Danish colony. Hunting and fishing issues were a major focus of these councils. In fact, as early as 1907, Greenlanders in Sisimiut proposed an international agreement for conserving both seals and whales.

When the Danish government signed the original 1931 whaling convention, however, it did so apparently without considering Greenland's interest. Neither the convention itself nor the Danish laws implementing it were placed before the councils for a hearing. When Denmark had the opportunity to express reservations about the convention before signing in 1934, it chose not to do so even though other countries (such as the United Kingdom) excluded all colonies from the convention's provisions before signing. In 1932, colonial authorities did present a regulation to the councils calling for a "prohibition against ruthless exploitation of baleen whales," however leaders of the two councils (both Danish officials) commented that the regulation would have little significance for Greenland. Subsequent implementation in Denmark of the 1931 convention was achieved under a royal decree that did not even apply to Greenland. Until the late 1950s, the Danish fisheries ministry reportedly was not even aware that the issue had any relevance to Greenland.

In 1950, Denmark became a signatory to the ICRW. As before, the provincial councils were never asked to review the action, despite the convention's language about aboriginal whaling. By 1959, however, Greenland's council (the two councils were merged in 1958) recognized that the country's longstanding humpback whale hunt was in apparent violation of the ICRW. It asked the Danish government to seek an exemption for this hunt, which the IWC granted in 1961.

In 1963, the fisheries ministry discovered that Danish laws and decrees implementing the convention did not apply to Greenland and proposed a new royal decree. The Greenlandic council supported the idea, but it rejected a draft because council members had "major objections to nearly all of the proposed stipulations." As one council report noted, "those who have written this draft have never participated in hunting. . . ." The leader of the council's debate wrote that, "although Denmark has taken on certain obligations because of the convention, the *landsråd* has in any case not been heard from earlier about this, just as there has not been any law applying it to Greenland; and any such law must be placed before the *landsråd*."[32]

This conflict between Danish and Greenlandic interests was placed before the Greenlandic Law Committee (*Det grønlandske lovudvalg*), which, under the Danish constitution, is charged with ensuring compatibility between Danish and Greenlandic law. The committee ruled that Denmark had obligated itself to carry out the provisions of the convention and that Greenlandic whaling was indeed covered. Brøsted's investigation of the committee's deliberations leads him to believe that it was ill informed when making this decision. He cites minutes of its meetings suggesting considerable uncertainty about the impending decision, with frequent use of phrases like "presumably applies . . ." and "regardless of the lack of clarity, one must presume. . . ." The committee's uncertainty was apparently resolved, according to Brøsted, when officials from the fisheries ministry and the Ministry for Greenland clarified several points. As a result, in 1963 the fisheries ministry issued a decree about whaling that was based upon a law that had questionable applicability to Greenland.

Greenland's marginalization by administrative action is also reflected in the fisheries ministry's translation of the ICRW provisions relating to aboriginal subsistence whaling. Following Danish approval of the ICRW in 1950, both the fisheries and foreign ministries accurately translated ICRW language about "aboriginal" whaling as that undertaken "for consumption by the aborigines" (Danish, *til de indfødtes forbrug*). When the fisheries ministry and Denmark's IWC commissioner took up the issue again in the late 1950s, however, the translation was altered to mean "for *local* consumption" (emphasis added), thereby removing any language referring to indigenous peoples. According to Brøsted,

the commissioner has acknowledged that the choice of words isn't perhaps the best, but that he doesn't like the word *aborigines* because it can have a connotation of racism . . . [and thus] the change is not a translation error but a change in meaning. Its use is so consistent that it can only be understood if seen in relation to the political situation and to Danish desires for Greenland. We now find ourselves in a time where Greenlandic policies are based on "normalization", equality and privatization, where Danes and Greenlanders should have identical rights and where the state's monopoly in commercial affairs should be reduced.[33]

A draft regulation about whaling, discussed with the provincial council in 1963, sought to build on this notion of "local" versus "aboriginal/indigenous" use. The draft read in part: "the right to carry out small whale hunting [here meaning humpback and minke whaling] is limited to Danish citizens with fixed residence in Denmark [including

Greenland], along with Danish citizens . . . who are given special permission." As Brøsted notes:

the language problem cannot be passed off as a translation mistake, because the concepts involved are different. All peoples can be viewed as local people, but only some people are the indigenous people of a land, who have lost control over their lands because of colonization by foreigners. The clear effect of the new choice of words was to do away with the notion that Greenlanders had any special right over Danes to catch whales in Greenland.[34]

This interpretation of the 1946 convention brought undesired consequences for Greenland. Denmark apparently succeeded in convincing the IWC that Greenlanders were not truly an indigenous people but were a "genetically and ethnically mixed" population. When Denmark asked the IWC in 1960 if Greenlandic whaling with fishing vessels and harpoon cannons was to be fully subject to the ICRW, the IWC thus responded affirmatively. In 1962, when the Danish commissioner sought IWC approval of a proposal to extend to "local" people the opportunity to whale for subsistence, the U.S. commissioner rose to state that "as chairman of the international conference that wrote the 1946 convention, I will say that the word aboriginal does not encompass any other people than indigenous residents, Eskimos, northern Indians, and Chukchis. This was the original idea." In 1963, Denmark dropped it proposal for lack of support from other member nations. Significantly, none of this debate nor the strategy behind it were discussed with Greenland's provincial council.

Greenlandic whaling under Home Rule, 1979 to the present

When Greenland achieved Home Rule in 1979, one of its first actions was to take over management of hunting and fishing;[35] yet, the Home Rule Act specifies that Greenland is subject to all international commitments binding on Denmark. With regard to whaling, Greenland is bound to abide by the regulations and quotas established by the IWC. It does so by working through the Danish Foreign Ministry and the Danish IWC delegation. As such, it must coordinate its efforts both with ministry officials and with representatives of the Faroe Islands, which also has Home Rule status.

An overview of IWC meetings in the 1980s and early 1990s reveals Greenland's difficult position in this international forum. In the late 1970s and early 1980s, Alaska bowhead whaling largely set the agenda for consideration of aboriginal subsistence whaling. However, it was also at this time that the IWC began to set quotas for Greenland's minke and fin whale hunts. During the IWC's 1980 annual meeting, both the Scientific and Technical Committees approved a limited bowhead hunt in Alaska following intense negotiations. This decision built on agreements reached in 1977 to allow a limited Alaska bowhead hunt, which Denmark also supported. At the same meeting, Greenland's humpback whale hunt also came up. Greenland had long had a fixed quota of 10 humpbacks in the IWC's schedule. But in 1981, the Scientific and Technical Committees recommended a zero quota. The IWC plenum overrode these recommendations (with support from Sweden and the United States) and allowed the hunt to continue. The difficulties in addressing these issues, however, led delegates to plan a meeting in 1981 of an ad hoc subcommittee on aboriginal subsistence whaling. It was agreed that both Greenland and the Inuit Circumpolar Conference (ICC) would be involved.

At the IWC's 1981 meeting, the ad hoc subcommittee presented its report entitled "Management principles and guidelines for subsistence catches of whales by indigenous (aboriginal) peoples." Representatives from the Home Rule government were present throughout the subcommittee's discussions. This report laid the groundwork for refinements to the aboriginal subsistence whaling regime in 1982, but the most immediate issue affecting Greenland at this meeting was that of quota infractions. The Technical Committee noted "with great seriousness" violations of these quotas and emphasized the need for greater control. The Danish delegate expressed hope that this problem could be resolved by the next meeting.

The IWC's 1982 annual meeting proved to be a major turning point both for Greenlandic whaling and for management of aboriginal whaling generally. Although the principal agenda item was the moratorium on commercial whaling, the commission also institutionalized procedures for addressing aboriginal whaling. In these discussions, Denmark sought to ensure that Greenlandic whaling was clearly recognized as aboriginal and not commercial whaling. Denmark also drafted special language amending the Schedule to reflect the new aboriginal whaling scheme. This proposal met opposition from Norway and the United Kingdom, which proposed a three-year "sunset" clause where Greenlandic and other aboriginal quotas would expire just as the moratorium went into effect. Denmark pressed successfully, however, for a comprehensive assessment of whale stocks by 1990 that would include revisiting the issue of aboriginal whaling. This proposal received the necessary three-quarters majority vote from the IWC plenum.

These conflicts in the IWC coincided with a pressing issue in another arena: Greenlanders' referendum on withdrawal from the European Community in 1982. Though Greenlanders' primary concern in the referendum was control over fisheries, they also feared outside control over marine mammal hunting. In previous years, proposals had surfaced suggesting that EC members speak with one voice in the IWC. Remembering well the support given in the 1970s to anti-sealing campaigns by some EC nations, Greenlanders feared being overwhelmed politically on an issue vital to their economy. As a result, they began to work more closely with Alaskan and Canadian Inuit on whaling issues. There was even discussion about creating an Inuit whaling commission, although this idea was placed on hold.

At the 1983 meeting, the Alaska bowhead issue continued to dominate aboriginal whaling discussions. A Mexican resolution to stop all aboriginal whaling on protected stocks (e.g., Alaska bowheads and Greenlandic humpback whales) failed, with eleven voting for and eleven against the resolution. The Scientific Committee again recommended that Greenland's humpback quota be zero; however, it said that if the plenum chose not to accept this recommendation, the quota should be four animals. The Technical Committee supported the Scientific Committee's proposal. In the end, Denmark was forced to accept a 10 percent reduction (from ten to nine) in its humpback catch. The Scientific Committee also recommended changes in Greenland's minke quota, taking an average of the previous three years' catch. This change was accepted by the Danish delegation and adopted by the plenum.

Greenland's use of the so-called "cold" harpoon for minke and fin whales also became an issue at this time. Concerns about "humane killing" led the IWC to ban the cold harpoon in commercial whaling as of 1982. Both Japan and Norway had since developed and used the effective penthrite exploding grenade. The United Kingdom and other member nations were now pushing for Greenland and other countries to begin using this new technology.

At the 1984 IWC meeting, delegates focused again on humpback quotas and repeated infractions. Denmark received a favorable response to its proposal to reduce humpback catches by another whale (nine to eight) in light of new concerns about stock status. To make up for this loss, the delegation proposed increasing the fin whale quota by two. Favorable action on this proposal, however, was made contingent on Greenland resolving the problem of repeated infractions.

The 1985 IWC meeting was, in the words of several Greenlandic participants, a "catastrophe." Greenland's quotas unexpectedly became a major focus of debate. Repeated infractions of humpback quotas clearly worked against Greenland's interests. In 1984, hunters had taken fifteen humpbacks, six more than the quota allowed. The Danish delegation was also apparently caught off-guard by a new statistical model for northeast Atlantic minke stocks that cast doubt on North Atlantic minke data. The delegation had to settle for less than half the earlier minke quota (a reduction from 300 to 130). It also lost the humpback quota entirely. One member of the Danish delegation re-called the meeting as "20 percent science, 80 percent politics," because Greenland was swept up in efforts by whale preservation groups to se-cure protection status for North Atlantic whale stocks.

The dramatic quota reductions imposed at the 1985 meeting also revealed tensions within Greenland between Home Rule authorities and KNAPK, the Greenlandic hunters' and fishers' association. Nikolaj Heinrich, then chairman of KNAPK, stated in the Greenlandic press:

The decision taken at the IWC's recent meeting in England that Greenland's whale hunt shall be cut in half, and moreover that the hunt for one whale species [humpbacks] be brought to an end, is absolutely not acceptable to KNAPK for the following reasons:
1) Greenland's population absolutely cannot give up whale meat.
2) whale hunting has great economic significance for Greenland's popu-lation.
3) anyone with knowledge of whales or who follows these issues cannot help but recognize that the stock of whales, and especially the larger whales, has clearly grown in recent years. In particular, there is certainly no basis for halting the hunt of humpback whales.

The demands being placed on Greenlandic whaling from the IWC, and with the support from both Danish and Home Rule authorities, cannot be accepted by KNAPK. We will go so far as to urge Denmark to leave the IWC, because this only increases our difficulties.[36]

These tensions spotlight the fledgling Home Rule government's ability to regulate and monitor whaling. As Brøsted's account makes clear, the application of IWC quotas and regulations to Greenland originally had been a matter of some dispute; however, as events pro-gressed, the Ministry for Greenland developed more comprehensive regulations based upon the IWC decisions. Royal decrees about whaling, issued first in 1939 and then in 1963, 1967, and 1974, prescribed the broad framework for Danish compliance with international whaling conventions. Decrees from 1963 on refer explicitly to Greenland as well as to Denmark. These statements specified which whales could be taken and what regulations governed both hunting and reporting of catches. In 1974, Greenland's provincial council also adopted regulations governing the catch of minke whales. It was this administrative history that the Home Rule government inherited in 1980.

In response to these tensions, Home Rule officials began adopting a more comprehensive approach to whaling management. They expanded on earlier efforts by enacting regulations specifying who can catch whales in Greenland. Regulations adopted in this period state that hunters must have "strong affiliation with Greenlandic society and have had permanent domicile in Greenland within the last two years."[37] Those involved in whaling must furthermore hold a full-time fishing and hunting license. To qualify for this license today, applicants have to demonstrate that they participated in hunting and fishing for at least 125 days in the previous calendar year or have not had a regular wage income in the previous 125 days.[38] The intent of these regulations is to limit whaling to those with adequate experience and expertise. In practice, it also virtually guarantees that only ethnic Greenlanders (and not Danes living in Greenland) can catch minke and fin whales.

In 1985, Home Rule officials also began an on-going effort to devise an effective and equitable quota allocation process. At first, they established a system of regional quotas for fin and humpback whales. This system was later modified after the regional quotas proved difficult to monitor and when humpback quotas were set at zero. Officials shifted instead to municipal quotas, with local councils taking on greater responsibility for hunt licensing and monitoring. They expanded this role to include minke quotas as well as 1987. This approach served to bring decisionmaking about whaling closer to the municipal level and to hunters themselves. It also provided more flexibility by enabling municipal councils to establish their own local criteria for approving permits and for allocating quotas between types of hunters. Municipal officials could now draw more upon the knowledge and experience of local hunters. While making this shift, the Home Rule also implemented for the first time a system whereby hunting violations would lead to quota reductions at the municipal level in the following year. The new approach thus brought municipalities not only greater involvement in decisionmaking but also more responsibilities for monitoring.

In 1986, the Home Rule responded to growing IWC concern about collective whaling. It adopted more stringent regulations, specifying under what conditions hunters can carry out this strategy. Basically, the collective hunt was allowed only with special dispensation, to be given only when:

1) the request is approved both by the local municipality and by the local hunters' and fishers' association;
2) it is documented that whaling has "major economic significance" for the local society in which the hunters live; and
3) that the need for fresh whale meat is so great that it can't be met by local fishing vessels outfitted with harpoon cannons.[39]

When dispensation is given, regulations specify that at least five skiffs must take part in a collective hunt to maximize chances of success. They also state that rifles of 7.62-millimeter or 9-millimeter caliber must be used (shotguns and fully automatic rifles are prohibited). All skiffs must be further outfitted with a hand-thrown harpoon and a float attached to a line at least twelve millimeters thick. The harpoon is to be attached to the whale before it dies to ensure that it doesn't sink. Rules state that the hunt must be led by a "captain" (Greenlandic, *aqut-toq*; literally, "helmsman") responsible for organizing the catch and reporting it to authorities. Hunters must also use the quickest means possible for killing the whale.

At the 1986 IWC meeting, the Danish delegation was again on the defensive as the Dutch commissioner raised concerns about a 1984 incident in which whale products were reportedly shipped from Greenland to Denmark and sold in a Copenhagen shop. The issue had been raised earlier in the Danish parliament, where Home Rule representatives stated that the export was contrary to Greenlandic law and was an unfortunate mistake. They promised the incident would not be repeated. The other major issue affecting Greenland at the meeting had to do with the collective hunt for minkes in East Greenland. Denmark was successful at this meeting in arguing that this hunt—like minke whaling in West Greenland—was aboriginal whaling. This brought consistency to the way in which the IWC addressed Greenlandic quotas.

Between 1985 and 1987, Denmark sought to turn around what some viewed as a deteriorating relationship with the IWC by expanding information provided to the IWC and by making efforts to ensure indigenous Greenlandic representation in its delegation. This initiative renewed efforts to insure that hunters themselves were directly represented on the delegation through KNAPK, the Greenland hunters' and fishers' association. KNAPK's role was spelled out in a letter sent to all IWC delegations:

1) to make sure that the needs and interests of the local population in Greenland are assured under aboriginal subsistence whaling;
2) to cooperate in improvements in hunting methods, including reduction of struck and lost rates, e.g. through the use of the penthrite harpoon-grenade . . . ;
3) to cooperate with scientists in counting whales and in collecting more knowledge about whale species in Greenlandic waters; and
4) to make sure that decisions taken by the IWC meet the subsistence and nutritional needs of the local population in Greenland.[40]

By including KNAPK in the delegation, Denmark sought to expand hunters' understanding of the IWC process and to work toward greater consensus within Greenland about future management policies. In 1986, there was also a change in the leadership of the Danish delegation; Henrik Fischer became commissioner, replacing Einar Lemche, who had taken a new position in the Home Rule government.

Despite these efforts to improve relationships, the IWC dramatically reduced Greenland's minke quota at its 1988 meeting, allocating only sixty whales (down from 110) to hunters in West Greenland and twelve to those in East Greenland. It did so in light of new information about the status of West Greenland's "stock" of minke whales. The IWC compensated for the reduction by approving a significant increase in the fin whale quota, raising it from ten to twenty-three. Although this move made up for the lost tonnage of minke products, it also altered the character of whaling. Fin whales can only be taken by vessels with harpoon cannons, whereas minkes can be taken either by vessels or in a collective hunt; thus the reduction in the minke quotas hit collective hunters particularly hard. Greenlanders on the Danish delegation—especially from KNAPK—faced sharp criticism at home because of this action.

At its 1989 meeting, the IWC increased the minke quota somewhat by approving a two-year (1990–91) quota of 190 minkes, with no more than 100 to be taken in one year. During the meeting, Denmark also presented a report on Greenlandic whaling that sought to provide a more comprehensive picture of the significance of whaling in contemporary Greenlandic society.[41]

In 1990 and 1991, the Home Rule government continued to refine its regulations for minke and fin whaling. One of the most significant changes was a requirement that hunters obtain an individual whaling permit in addition to their normal fishing and hunting license. The Home Rule delegated to municipalities the authority to issue these permits. For the collective hunt, a hunt "captain" is selected to receive the permit and is responsible for reporting the catch. In addition, hunters wishing to sell whale products are now required to have their permit stamped by municipal authorities to ensure accurate reporting of whales taken.

Another major change was a ban on use of the "cold" (nonexploding) harpoon, long a source of controversy in the IWC. Beginning in April of 1991, all vessel hunters were required to use a fifty-millimeter harpoon cannon with an exploding penthrite grenade. This development came about as a result of a cooperative program begun in 1987 involving Home Rule officials, KNAPK, and vessel owners.[42] In 1990, the Home Rule purchased one hundred of the penthrite grenades from Norway on a trial basis. Seventy of these were designed for use on minke whales, and thirty, for fin whales. Of the one hundred grenades, fifteen were given to KNAPK for tests by selected vessel whalers in Ilulissat, Aasiaat, Maniitsoq, and Nuuk. KNAPK prepared a videotape in Greenlandic describing the program and explaining how the grenades are used. After successful tests, the remaining grenades were made available in 1991 to hunters who had successfully completed a KNAPK course about their safe and effective use. With the new regulations in place, all vessel whalers were required to participate in further training courses held in 1992 and 1994 before they are allowed to purchase the grenades.

In 1992, the Home Rule also implemented a program for renovation of harpoon cannons, many of which were in poor condition and dangerous to use. The cost of these renovations ranged from about $10,000 to as much as $27,000 (60,000–160,000 DKK). Because this cost was far beyond the means of most vessel owners, the Home Rule itself allocated over $800,000 ($5 million DKK) to help repair over sixty cannons. Vessel owners themselves paid one-third of the costs. The result of this effort is that harpoon cannons are now safer and more accurate, thereby reducing risks to hunters and improving the efficiency of the hunt.

More recently, additional whaling regulations have been enacted by the Home Rule government. Table 15 shows selected regulations in effect for minke and fin whaling in 1995. A major change made in 1993 requires for the first time that minke whales struck but lost in West Greenland be counted against the quota. Previously, whales struck but

TABLE 15.
Selected Greenlandic whaling regulations, 1995

Regulation	Minke whaling	Fin whaling
Type of hunt	Vessel whaling *or* Collective whaling (only with special dispensation)	Vessel whaling only
Hunter licensing	Full-time hunting license Permanent resident Close affiliation with Greenlandic society	Full-time hunting license Permanent resident Close affiliation with Greenlandic society
Whaling license	Required from municipality Dispensation for collective hunt when hunt has major significance for local community and where meat from vessel whaling not available	Required from municipality Can be issued to: 1 vessel ≥36′ long 2 vessels ≥30′ long
Season	1 April–31 December	1 January–31 December
Hunt requirements	No females with young may be taken Whale must be killed as quickly as possible Use of vessel with harpoon cannon (≥50mm) Harpoon grenade required Training for harpooner Cannon in good condition Registration & inspection of cannon by authorities Equipped with winch *or* Special dispensation for collective hunt Minimum of 5 skiffs Use 7.62mm rifles or larger Full automatic rifles prohibited All skiffs equipped w/hand harpoon, float & ≥12mm line Designated hunt leader All edible meat and *mattak* must be used Catch data must be reported to authorities before any sale of whale products Sample of whale meat & *mattak* must be provided for research	No females with young may be taken Whale must be killed as quickly as possible Use of vessel with harpoon cannon (≥50mm) Harpoon grenade required Training for harpooner Cannon in good condition Registration & inspection of cannon by authorities Equipped with winch Whale must be ≥15.2m All edible meat and *mattak* must be used Catch data must be reported to authorities before any sale of whale products Sample of whale meat & *mattak* must be provided for research

SOURCE: Greenland Home Rule Government.

not retrieved were not included in catch statistics. A system for reallocating unused quotas has also been developed. Any quotas unused because of poor weather conditions, problems with equipment, or other reasons can be reallocated in the fall of each year to municipalities in the ice-free areas from Sisimiut south.

As this discussion shows, Home Rule whaling regulations today are the result of more than a decade of progressive improvements and refinements. Importantly, they also reflect a process of mutual education—education of hunters about changing national and international expectations for whaling, and of "outsiders" about Greenlandic whaling traditions and needs. As Greenland's minister for fisheries and industry put it recently:

In Greenland, we . . . have to make a constant effort to ensure that our hunting meets standards of sustainable utilization, and that we can document this for the outside world. . . . [Thus] Greenland works domestically for:
- the best possible distribution and utilization of hunting potential;
- as accurate reporting of catches as possible;
- that the necessary resources are made available for the development and improvement of hunting equipment;
- intervention if hunting regulations are infringed;
- that the necessary resources are made available for biological research.[43]

Today, quota allocations to municipalities are approved by the *Landstyre* in consultation with KNAPK and KANUKOKA. Table 16 shows these allocations for 1993 and 1994, as well as the allocation between vessels and collective hunts. They are based upon four criteria. The first is population: Officials estimate approximately how much meat, *mattak*, and other whale products can be expected from the entire national quota and then make a preliminary allocation on an equal per capita basis. They also determine whether this allocation should come from minke whales or fin whales (or both) based upon the whales' seasonal availability. They then adjust the initial allocation based on the second criterion—the number of vessels with harpoon cannons in a given municipality. This number helps determine to what extent the municipality will receive minke or fin whales. Those with more vessels and cannons are likely to receive a higher fin whale quota. A third criterion is the availability of alternative sources of income in the municipality. Those with fewer opportunities may get a higher quota. Finally, the total number of settlements in a municipality is taken into account. If there are many settlements, quotas for minke whales may be set higher to ensure widespread distribution of hunting opportunities. Once quo-

TABLE 16.

Allocation of minke whale quotas and catches in 1993 and 1994, by municipality

	1993		1994	
Municipality	Quota total (vessel/ collective)	Catch (landed + struck + other)	Quota total (vessel/ collective)	Catch (landed + struck + other)
Nanortalik	9 (4/5)	9 + 1 + 1	10 (4/6)	11 + 0 + 0
Qaqortoq	9 (4/5)	8 + 1 + 0	8 (3/5)	8 + 0 + 0
Narsaq	5 (3/2)	5 + 0 + 0	5 (3/2)	6 + 0 + 0
Paamiut	8 (6/2)	10 + 0 + 0	8 (6/2)	8 + 0 + 0
Nuuk	15 (14/1)	17 + 0 + 0	15 (13/2)	16 + 0 + 0
Maniitsoq	12 (9/3)	13 + 0 + 2	12 (9/3)	13 + 0 + 0
Sisimiut	13 (12/1)	13 + 1 + 2	13 (12/1)	15 + 1 + 1
Kangaatsiaq	5 (3/2)	5 + 1 + 0	3 (1/2)	3 + 0 + 1
Aasiaat	8 (4/4)	7 + 0 + 0	8 (4/4)	6 + 0 + 1
Qasigiannguit	5 (3/2)	4 + 1 + 0	5 (3/2)	5 + 1 + 0
Ilulissat	7 (6/1)	4 + 0 + 0	7 (6/1)	7 + 0 + 0
Qeqertarsuaq	5 (3/2)	4 + 0 + 1	5 (3/2)	4 + 1 + 0
Uummannaq	3 (0/3)	1 + 1 + 0	3 (0/3)	1 + 0 + 0
Upernavik	3 (0/3)	1 + 0 + 0	3 (0/3)	1 + 0 + 0
Totals/West Greenland	107 (70/37)	101 + 6 + 6	105 (67/38)	98 + 4 + 2
Total/East Greenland	12	9	12	5

SOURCE: Greenland Home Rule Government.

tas are allocated, municipal councils work with local hunters' and fishers' associations to determine which individuals will receive a permit. These permits are valid for the calendar year.

Current regulations state that fin whales can be taken throughout the year in West Greenland. They must be at least 15.2 meters in length, and females with young may not be taken. As in vessel whaling for minkes, fin whales may only be hunted with a fifty-millimeter harpoon cannon and exploding grenade. The harpoon must be attached to a line (at least twenty millimeters in diameter) and a large float. This line helps ensure that a whale is not lost. Municipal authorities can issue a permit to two smaller vessels should they desire to work together in the hunt, or to a single vessel if it is at least thirty-six feet in length. All vessels involved in fin whaling must have a hydraulic winch to assist in processing.

In West Greenland, minke whales may only be caught from 1 April to 31 December; however, if a minke is found in a *sassat* (an entrapment of whales by ice), it can be taken at any time. Females with young may not be taken. Vessels used in minke whaling must also use the fifty-millimeter harpoon cannon and must have a power winch. If the exploding grenade fails to kill the whale immediately, hunters may use another grenade or rifles at least 7.62 millimeters in size. Fully automatic rifles are not allowed. For the collective minke hunt, permits require that at least five skiffs participate, that hunters use rifles at least 7.62 millimeters in size, and that a harpoon with a twelve-millimeter line and a float be used.

Permit reporting requirements are similar for both minke and fin whaling. All who participate in whaling have an obligation to report fully about the hunt and about the whale taken. In West Greenland, a minke permit is considered used once a whale has been struck, either with a harpoon cannon or with a rifle. To ensure accurate reporting, whale products cannot be sold before municipal authorities have stamped the permit. Hunters are required to use all edible portions of the whale. If for any reason the entire portion cannot be used, hunters have an obligation to give the remaining portion to the local population. Hunters must also submit a small piece of meat and *mattak* from their catch for biological research. Those who violate any of these regulations can face both fines and confiscation of all whale products. Home Rule authorities retain the right to invalidate permits or to reallocate them to other municipalities after consultation with KNAPK and KANUKOKA.

Costs of co-management:
Whaling and social differentiation in Greenland

In this and subsequent sections, I highlight pressing controversies surrounding Greenland's whaling regime today. The first part focuses on social differentiation and the costs of co-management—conflicts generated within Greenlandic society by expanded involvement in management and monitoring. Later sections address conflicts over the validity of scientific data in managing whaling, over the use of new hunting technologies, and about the morality of selling whale products locally for cash.

As the narrative in chapter 1 shows, the issue of Greenlandic identity was central to conflicts leading to Home Rule. Language, occupation, food, kinship, ways of sharing, ways of thinking—all have been used over time as measures of who is a "real" Greenlander. To some extent, this debate continues today within Greenlandic society. For better or worse, the whaling issue forces this debate into the public policy arena. It touches a sensitive nerve by highlighting increasing social differentiation in a society that prefers to think of itself as egalitarian, and it causes many Greenlanders to ponder the multitude of changes affecting their society.

When the IWC reduced minke whale quotas by over half in the late 1980s, submerged tensions between fishing vessel owners and collective hunters came to the surface. Vessel owners are often perceived as "wealthy" by Greenlandic standards. Collective hunters, in contrast, are typically less well off; in some cases, they represent the poorest group in local economies. The collective hunt is often most visible in smaller settlements where there are few economic opportunities. These tensions thus add to on-going debates about Greenlandic identity and values and how best to balance development between larger towns and smaller settlements. Conflicts between these groups over whaling regulations were perceived by some as favoring vessel owners over collective hunters. For a time, this tension led one prominent Greenlander to express fears about an emerging internal "class war," with each group vying for their share of whale quotas. The conflict was aggravated by Home Rule regulations that allowed only those holding full-time fishing and hunting licenses to whale. At the time, these regulations were even more stringent than they are today. The result was that, with reduced quotas—especially for the collective hunt—many experienced hunters were excluded from whaling. The small quotas allocated by the IWC in the late 1980s created special hardships and bitterness among many collective hunters:

The collective hunt is the only way to get a little meat and *mattak* now. The problem is, because the quota is so low, many hunters go out after one whale. It is dangerous to have so many people out here shooting. Last winter in a place not far from here, one hunter was wounded. . . . Before the quotas, there would usually be six or eight or ten boats going out after one whale. Others knew that they'd have a chance later on. But now everyone goes out because maybe we can only get a few whales each year.[44]

Others spoke about unfriendly competition over who would be included in distributions from a collective hunt. One disappointed hunter recalled showing up too late to receive any meat or *mattak*: "During the last hunt, we arrived too late to participate (*inortuigatta*). The hunters who spotted the whale only let their family members know about it, and we were excluded. It's all because of these quotas."[45]

This frustration created a political atmosphere that some feared could lead hunters to reject any outside management or control. In fact, the quota violations of the 1980s might be attributable to these tensions. The issue of infractions has been a continuing concern in the IWC. Of some sixty-seven local quotas allocated by the Home Rule to hunters in West Greenland between 1977 and 1990, nearly 30 percent have been exceeded.[46] However, these infractions rarely exceeded the overall IWC quota, and they have declined in recent years. But as one Greenlander said, speaking anonymously, "Greenlanders are a very law-abiding people. But if you push a regulation on them from the outside that makes no sense in their daily lives, you can create some very unpleasant results."

A few hunters have criticized the Home Rule government itself. For them,whaling regulations emanating from Nuuk seem little different than those imposed from a distant colonial capital. For some, Nuuk becomes simply another distant "center," issuing regulations to "peripheral" communities that don't mesh with day-to-day reality.

In the context of the settlements . . . general orders and regulations received from outside very easily come to be seen as irrelevant, if not downright incomprehensible. By accepting their jobs, the local [Greenlandic] heads of administration have taken it upon themselves to handle a sometimes difficult intermediate position by being, on the one hand, physically present and members of the local communities while, on the other hand, having to represent and—from time to time—enforce directives issued by far-away bodies.[47]

These Home Rule officials counter that Greenland, as an emerging nation, has an obligation to live up to its commitments to the international community. "If we ignore violations of whale quotas," stated Greenland's minister of fisheries and industry, "it will be interpreted as if we are indifferent to agreements we have committed ourselves to, and it will put Greenland in a very poor light."[48] Said another Greenlander working in Home Rule offices in Nuuk:

It isn't us here in [the Home Rule administration] who have imposed the things in the regulations that some hunters call ridiculous. Things like fin whales that may only be caught when they are over a certain size. Hunters say "What are we supposed to do? Pick the whale up and measure it . . . ?!" These regulations are not easy to adhere to, but it isn't us who decide them. They are guidelines laid down and decisions made by the IWC. . . . We here in Greenland have to live with the fact that the rest of the world judges us because we are involved with whaling. And we in this country also live with the fact that [the IWC's quotas] create a lot of dissatisfaction.

We should remember, that many nations are against all forms of whaling, and Greenland may only catch a certain number of whales under the quotas: minke and fin whales. . . . All of this can be undermined if protected whales are caught unlawfully. It is in the hunter's interest, and indeed in everyone's interest, that only those whales that are supposed to be caught, *are* caught.[49]

The issue of quota violations came into the spotlight once again January 1994 when ten hunters in Kangaatsiaq Municipality in central West Greenland killed a humpback whale—a protected species under IWC rules—that was caught in the ice near two small settlements. The IWC halted hunting for humpbacks in the mid-1980s because of concern about stock size. Hunters believed that the whale would soon die because the hole in which it was trapped was freezing shut, but because the catch violated Greenlandic law, Home Rule officials cut the municipality's whale quota in half for the following year, consistent with regulations. Said the Home Rule official in charge, "the rules are clear enough. The quota reduction is the penalty called for in the regulations governing the catch of large whales."[50] Some local politicians objected, wondering outloud, "who do [these officials] work for—Greenland's people or environmental organizations in other countries?"[51] Home Rule officials held fast, however. This action brought a positive response from environmental organizations, including Greenpeace, whose spokesperson stated:

It is important for Greenland to show the world that it takes whale protection regulations and the violation of them seriously. It is sensible for the Home Rule to react just as it has in reducing Kangaatsiaq Municipality's quota for 1994, because it shows the IWC that Greenland takes the international system seriously.[52]

Increases in IWC quotas since 1989—especially for minke whales—have eased tensions somewhat surrounding quotas and regulations. The Home Rule has taken steps to reduce inequities in licensing so that experienced hunters who work part-time can also participate in whaling. Some municipalities are also implementing their own criteria for allocating whaling licenses to vessels not involved in the shrimp fishery. Many hunters say that tensions could be reduced even more if the IWC increased the number of minkes available, even if it results in a reduced fin whale quota. This suggestion is made because minkes are smaller and easier to process and the hunt doesn't necessarily require a vessel. The quota change could enable more hunters to participate and lead to a broader distribution of whale products in local communities.

Another factor affecting Greenland's whaling regime is a frequent lack of consensus among hunters, biologists, and political leaders about the status of whale stocks. Too often, reports from the IWC that humpback whale and minke stocks are greatly reduced conflicts with hunters'

experience. Hunters and fishermen see whales frequently in Greenlandic waters and typically feel they have a better sense for whale population dynamics than do researchers attempting to count whales from
aircraft or boats. In a recent letter to the IWC, for example, KNAPK
took issue sharply with data presented to the Scientific Committee:

KNAPK is fully aware of the [whale survey] results presented in the Scientific
Committee, based on one year's research . . . , and would like to state the fact
that one year's research is not a representative basis in allowing further cutbacks on takes of minke whales. KNAPK cannot in any way agree on the scientific estimates of the minke stocks in West Greenland. . . . [I]t is the hunters'
and fishermen's strong impression that there are at least three times more
minke whales than the estimates from [the] Scientific Committee.[53]

KNAPK went on to describe its own members' efforts to estimate the
size of minke stocks in West Greenland. Another newspaper article appeared recently with the headline, "The ocean is full of whales." In the
article, hunters in Nuuk expressed the belief that there are "huge numbers of whales"—some say even "too many"—in Greenlandic waters.[54]
These local reports carry considerable weight in Greenland's small society, reliant as it is on oral traditions. Hunters' accounts of whale
sightings and hunts are heard frequently in Greenlandic radio broad-
casts and have a considerable impact on public opinion about whaling.

Home Rule officials take this conflict between regulation and perceptions seriously, although monitoring and enforcement of regulations is a difficult and sensitive subject. Officials are now informing
hunters about whaling issues and about what biologists believe is happening with local stocks. They are mounting an education campaign to
teach younger hunters to identify whales accurately. As part of that effort, they produced a large poster detailing field characteristics useful
for proper whale identification. They are also producing radio programs informing hunters about IWC actions and recent whale research.

In recent years, the Home Rule and local municipalities have gone
one step further by hiring fish and wildlife officers to monitor compliance with hunting and fishing regulations. Officers are currently located in Uummannaq, Ilulisaat, Sisimiut, Maniitsoq, and Nuuk. A major part of their job is educating the public about regulations and
reporting violations to the police. Hiring conservation officers may not
seem extraordinary in many countries, but in the Greenlandic context
it represents a new stage in resource management. Although the officers do not focus on whaling alone, their presence expands the visibility
of Home Rule and municipal resource management efforts.

The IWC itself is placing greater emphasis on monitoring and enforcement issues. As part of a broader discussion surrounding the RMP, a working group is exploring how international observers could be placed on board vessels involved in commercial whaling to ensure compliance with quotas and regulations. If this approach were ever attempted in Greenland, there would be many questions about how this could be done realistically, especially given the fact that whaling is done opportunistically. It would be extremely difficult to predict when an observer would have to be on board; furthermore, the vessels used in Greenland are small, and there is little extra room. There are also issues of language (most hunters speak only Greenlandic) and cost. These issues are amplified many times over in the context of collective whaling, where hunters typically live in remote communities spread along Greenland's extensive coastline.

Conflicting data:
Science and uncertainty in the management of whaling

This lack of consensus about the status of whale stocks in Greenland brings up the difficult issue of managing whaling in the face of uncertainty. Conflicting data and interpretations about stock identity, size, and productivity make effective management difficult. The tools available to marine mammal biologists are increasingly sophisticated, but logistical problems, poor weather, and high costs continue to make gathering data difficult.

At the same time that the IWC decided to implement a moratorium on commercial whaling, it also agreed to undertake a comprehensive assessment of whale stocks by 1990. As it turned out, that effort took several additional years and focused largely on a few, high-profile stocks such as minke whales in the northeast Atlantic. One of the goals of the comprehensive assessment was to review existing data about stock identity and size and to explore means of improving management. In the Greenlandic context, management of whaling has been hampered by uncertainty about stock identity and size. The IWC's Scientific Committee has mapped out boundaries for minke and fin whale stocks in the North Atlantic that are used in setting quotas (figure 19). For example, a line on the map drawn due south from Greenland's southernmost point (Cape Farewell) is used to distinguish a West Greenland

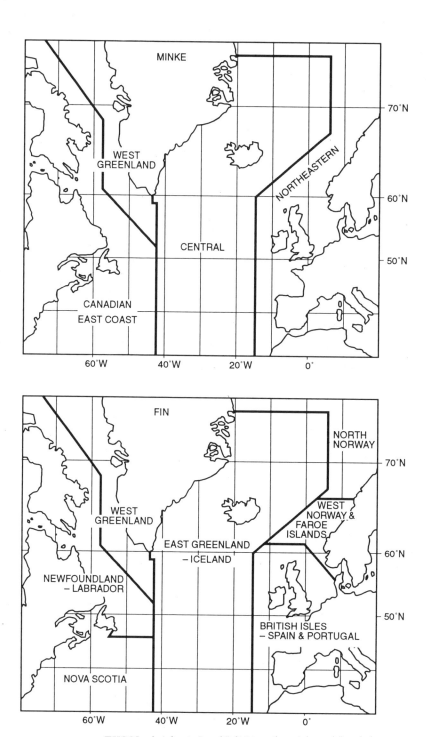

FIGURE 19. IWC North Atlantic "stock" divisions for minke and fin whales

"stock" of minkes and a central Atlantic "stock." In recent years, however, it has become clear that this line has little scientific basis. The West Greenland "stock" is very likely not separate at all but is almost certainly part of a larger stock. The Scientific Committee acknowledges this difficulty but is unable to change the boundary because it can't really say where it *should* be. This issue has important management implications. If Greenlanders are hunting from a small and discrete stock, then catches must be relatively small. If, however, the stock is actually much larger, quotas for sustainable use could perhaps be set higher. The Home Rule government has repeatedly stated the need for 670 tons of meat from larger whales each year to meet nutritional needs.[55] KNAPK believes that 800 tons annually are needed.[56] Current quotas provide for the equivalent of about 500 tons.[57]

Uncertainty is an inherent part of marine mammal management. In the Greenlandic context, biologists face the difficult task of counting whales from vessels or aircraft while being buffeted by winds and high seas. Cooperative international sighting surveys and new research methods, especially the electrophoretic technique (the so-called DNA fingerprinting) for determining stock identity, promise to reduce some of this uncertainty in the years ahead. Cost is clearly a factor in this type of research; Home Rule budget cutbacks place even greater demands on biologists responsible for these surveys. Scientific research is an essential part of effective co-management, however, and funding will have to be found if conflicts between hunters, biologists, and policymakers are to be resolved.

Conflicting ideologies:
Cash, commoditization, and the morality of exchange

In the IWC context, controversies over Greenlandic whaling focus not only on biological issues but also on ideological debates about the meaning of "subsistence," the role of cash in mixed economies, and the morality of exchanging whale products locally for cash. I argue in the introduction that these debates are rooted in conflicting idea-systems arising from differing modes of production. Defining "subsistence" has proven difficult in the North. "Subsistence is like pornography," some say, "everyone knows it when they see it, but no one can define it." Defining "subsistence whaling" in the IWC context is difficult because of ambiguity about the meaning of the words "commercial" and "subsistence."[58] Is it still "subsistence" when Greenlandic hunters sell whale meat and *mattak* for local consumption? Some find the sale of any whale products anathema; but views are often shaped by assumptions about the morality of exchange involving cash and about the effects of cash and commoditization on kin-ordered societies. Anthropologist Brian Moeran argues that distinctions between "commercial" and "subsistence" whaling are based on a false dichotomy because they ignore cultural constructions of value about money and exchange:

> . . . although money may seem to most Westerners to be impersonal and to signify a sphere of purely "economic" relationships, this should not be seen as being necessarily the case for other societies . . . where money can express relationships which are ideally personal (rather than impersonal), enduring (rather than transitory), moral (rather than amoral), and altruistic (rather than calculating). In other words, it is our mistake to see the "economic" as somehow divorced from other social relationships and forming an autonomous domain.[59]

In the Greenlandic context, Lynge echoes this argument by stating that sales involving cash are an extension of longstanding Inuit exchange relationships. In this view, it isn't the presence or absence of cash alone that distinguishes commercial from subsistence practices; rather, the cultural context of exchange relationships is what must be considered. In Euro-American society, "cold hard cash" is typically a means to accumulating wealth—profit maximization is the goal. In northern societies like that in Greenland, however, the role of money is more complex:

> The role of money in the modern hunting society has nothing to do with depreciation of investments or payment of crews. As far as such things are necessary, it is financed with the money that circulates through the fisheries. [But] seals and whales are hunted for their meat, and the meat that is landed is distributed to everyone on land who is interested. The distribution is performed along guidelines having their roots partly in the old rules about hunt shares and partly according to the fact that many people are not covered by these traditional regulations, but want to have some of the meat, and they have money. All hunters need cash to obtain daily necessities. Money is the only means of exchange that can open the channels of distribution in a modern society. The fact is that everybody wants meat, but only one-fifth of the population live in places where landing the meat is the general way of life. . . . Therefore it is sold. It goes for money.[60]

A second assumption centers on the impact of cash and commoditization on kin-ordered societies. Many assume that these forces inevitably undermine "subsistence"-based societies; however, the evidence from Qeqertarsuaq (in chapter 2) and elsewhere shows that mixed economies in the North can combine a cash sector with high levels of country food production. There are signs that the mixed subsistence-cash economies on the world's peripheries are perpetuating themselves over time. There is thus no significant, unambiguous effect of cash on these production relations. As Nicholas Peterson argues,

although cash and commoditization have been widely seen as inimical to the social relations taken to characterize small scale communities based on me-chanical solidarity, setting off a one way process leading to the destruction and replacement of indigenous practices and beliefs by those of the encapsulating society, it is increasingly clear that there is no single set of consequences.[61]

It is unwise to assume that the existence of local markets for whale products and the use of cash as a medium for exchange have the same meaning in Euro-American and kin-ordered societies. Greenlanders and other northern peoples are struggling to strike a balance between the demands of world markets and the security provided by kin-based systems of production. It remains to be seen whether this balance can be achieved over the long run, but Peterson believes that

if economic activity is socially constructed . . . then it is possible that as well as being transformed by these external influences foragers may assimilate some, many or all of the intrusions and linkages with the dominant economy to their own internal social purposes and in doing so reproduce distinctive sets of eco-nomic and social relations.[62]

From umiaq to hot harpoon: New technologies, new traditions

Conflicts over hunting technology in Greenland are illustrated clearly by a recent newspaper article with the headline, "The Eskimos Aren't Real Enough."[63] The article summarizes findings of a report produced by a major environmental organization asserting that in Greenland "the Eskimos . . . aren't real enough. They have speedboats, they no longer use kayaks, they no longer live like Stone Age people, they have videos and TVs." The source of the article is a well-known Icelandic critic of environmentalists, who has his own message to convey, but the quote illustrates nicely the constraints facing Greenlandic hunters in the IWC context due to conflicting views about technology and aborigi-nal whaling.

Greenlanders make no secret of their use of modern technology in whaling, including harpoon cannons, rifles, and outboard motors. They use this equipment because it is available, effective, and efficient. In their view, its use is in keeping with cultural traditions of flexibility and adaptability that have enabled Inuit to survive in the Arctic for generations. There is no permanent cultural attachment of Inuit hunters to any one form of technology. New technological traditions are created as the need arises.[64]

The creation of new technological traditions in whaling runs up against Euro-American stereotypes about Greenlandic society. However the persistence of these stereotypes is impressive. We all learn in school about "Eskimos" living in igloos, paddling kayaks, and eating raw meat; images made famous in Robert Flaherty's classic film, *Nanook of the North*, and in countless books about the Arctic. When we find out that Greenlandic and other Inuit societies have been changing along with our own, we often feel disappointed. Some even resent Greenlanders for this change, feeling that they are no longer "real enough."

Not surprisingly, then, in debates about whaling and technology, Greenlanders often feel caught in a dilemma. On the one hand, they are criticized for "not being Greenlandic enough." On the other hand, they are told to use the most effective (often the most technologically sophisticated) equipment to meet concerns about the humane killing of whales. Greenlanders are well aware that modern equipment like the penthrite grenade kills a whale much more efficiently than does a "cold" harpoon. As author Janne Jervin notes:

The time to death of the whales will definitely be reduced with the use of high-tech hunting gear . . . but the introduction of these high-tech weapons is in itself the source of a new problem: it clashes with traditional whaling. By responding to the call for more humane killing methods in subsistence whaling, whale hunters are immediately confronted with the problem stemming from the popular misconception that subsistence whaling should not be high-tech.[65]

Greenlandic hunters have adopted use of the penthrite grenade, even though the IWC has not formally insisted on it for aboriginal whaling. They have done so even though some believe the grenade causes undue waste of meat in minke whales;[66] nevertheless, in the IWC context many hunters feel damned if they use new technologies and damned if they don't.

This chapter illustrates the complex array of institutions, processes, and histories that make up Greenland's aboriginal subsistence whaling regime. Greenland's involvement with the IWC mirrors its growing incorporation into the world economy. Its hunting regime is responsive to external concerns and is increasingly effective in regulating and monitoring whaling. It is also making significant strides to improve whaling technology to address concerns both about crew safety and humane killing. Implementing this system has not always been easy, however. The Home Rule faces disgruntled hunters who resent reduced quotas and increasing regulation. It also continues to confront ideological challenges to whaling both within the Danish realm and within the IWC. As we shall see in chapter 5, recent events have increased conflicts in both of these arenas, forcing Greenlanders to take new political initiatives to protect their interests.

❧ 5 ❧

Initiatives on the Periphery: Home Rule and the Politics of Whaling

I n *Weapons of the weak: Everyday forms of peasant resistance*, anthropologist James Scott describes how individuals and societies on the world's peripheries are resisting being characterized simply as "victims of progress," incapable of countering colonialism's legacy.[1] Scott's book supports the view presented in the Introduction that peripheral societies may well have tools for empowerment available to them. Drawing upon fieldwork in Malaysia, Scott describes how peasants often practice "cautious resistance and calculated conformity" to undermine external authority and enhance self-determination. In some settings, peasants may simply feign compliance with a landlord's admonition, or they may drag their feet in implementing government programs. In other situations, they may openly defy official laws and regulations. According to Scott, peasants on the world's peripheries quietly carry out countless daily acts of resistance in the spirit of the Ethiopian proverb, "When the great lord passes, the wise peasant bows deeply and silently farts."[2]

Goldsmith and colleagues show how these same quiet strategies are increasingly employed in managing common property resources, from forests and pastures to wildlife and fisheries.[3] As the power of colonialism wanes, local resource users are finding new opportunities within faltering international economic and political systems preoccupied by recessions, burdensome debt, and the "balkanization" of empires. In some cases, these opportunities enable them to recapture common property resources and to reinvigorate local customs and practices. Many peoples, in fact, relish the prospect of systems eroding even further, ushering in more opportunities to define their own priorities and identities and to restore what colonialism has disrupted.

In this chapter, we explore two such acts of resistance and calculated conformity in the context of Greenlandic whaling. Far from being "passive victims" of external political forces, Greenlanders are demonstrating through the Home Rule process how careful thinking and shrewd political calculation can help overcome political marginalization in the IWC. The first such initiative involves participation in the North Atlantic Marine Mammals Commission (NAMMCO), a regional research and management organization for marine mammals. As a founding member of NAMMCO, Greenland joins the Faroe Islands, Iceland, and Norway in building a resource management institution that could perhaps become an alternative to the IWC. In a second parallel initiative, Greenland has become actively involved with the Inuit Circumpolar Conference (ICC) on whaling issues. At the ICC's 1995 meeting in Alaska, participants discussed a new "ICC Whaling Agenda." Among other options, the ICC discussed creating a new "Inuit circumpolar whaling commission."

Just how these two Greenlandic initiatives develop remains to be seen. Significantly, although Greenland pursues them, Home Rule officials insist that the country will remain a part of the Danish delegation at the IWC. These dual initiatives suggest, however, that Greenland may be practicing "cautious resistance and calculated conformity" on the international level with regard to whaling. At the very least, it demonstrates that Greenland does not intend to be passive when it comes to the politics of marine mammal management.

Before examining these initiatives in detail, we need to look at two ideological debates that inform them. The first is the growing conflict at the international level over animal rights. The second has to do with conflict within the Danish realm over whaling policies. Both have significant impact on the way Greenlanders think about whaling issues, and they influence the options available to the Home Rule government for responding at the international level.

Of seals, small cetaceans, and animal rights: The ethics of killing marine mammals

Greenland's political initiatives arise from its history of confrontation with environmental and animal rights groups inside and outside of the IWC over the ethics of killing marine mammals. A full account of this history is beyond the scope of this study,[4] but the confrontations began in earnest during the 1970s, when certain groups, including

Greenpeace and the International Fund for Animal Welfare (IFAW), undertook a highly publicized campaign to halt commercial seal catches off Newfoundland. This effort culminated in the 1983 decision by the European Economic Community (recently renewed) to ban all imports of sealskins. From the perspective of animal rights groups, it was, in Wenzel's words, "a spectacular victory. . . . No environmental, or animal rights campaign has since had such sweeping success."[5]

Anti-sealing activists failed, however, to distinguish between intensive, ship-based sealing off Newfoundland and community-based sealing in Canada and Greenland. In the public's mind, all were linked together. The subsequent ban on sealskin purchases had a devastating impact on Greenlandic livelihoods. According to Wenzel, the efforts of Home Rule officials to convince these groups of their wrongheadedness had limited effect.[6] In the 1980s, the positions of animal rights organizations hardened. Conflicts over sealing expanded into broad anti-hunting and anti-trapping crusades that challenged the very basis of Inuit society. The comments of Paul Watson, leader of the Sea Shepherd Society, illustrate vividly the ideological nature of this conflict. Speaking about the impact of the anti-harvest campaign on indigenous peoples in the North, Watson argues that

this is an era of changing social values. . . . [T]raditions will be broken, people on both sides will be hurt but it is a part of natural human evolution. . . . [S]hould social change within the context of one social group be restricted by the result it will have on another social group, especially in light of the fact that such social change is perceived . . . as being progressive?[7]

Watson argues that the dominance of European ideologies about the environment over those of Greenlanders and other indigenous peoples is an inevitable outcome of "natural human evolution." The reason for this dominance, he argues, is that at heart all people are driven by the same instinct: profit. Steven Best, a spokesperson for the International Wildlife Coalition, makes a similar argument: "All people are essentially the same. . . . Inuit kill for the same reasons as Newfoundlanders, as South Africans, as Scottish fishermen, to make some money. They may believe otherwise, but everyone believes otherwise."[8] According to this argument, making money is the primary motivation behind all resource use, no matter who does it, what its significance is locally, or what scale it is conducted on.

Another argument that Greenlanders confront is that whales possess an inherent "right to life." In a recent article, for example, American attorneys D'Amato and Chopra argue that whales—because of their presumed intelligence—have such a "right to life." In their view, the cultural rights of Greenlanders to their subsistence foods cannot stand: "The Inuit's claims are at the expense of an overlooked voice—the anguished cry—of the sentient inhabitants of the deep. . . . The whales find their own sustenance in the oceans; by what right do the Inuit expropriate the bodies of the whales to serve as their food?"[9]

These views, though extreme, are not isolated aberrations. Increasingly, they appear to be part of a strategy to deny Greenlanders their history as hunters, and the debate is now widening into controversies over management of small cetaceans, notably beluga (*Delphinapterus leucas*, or white whale) and narwhal (*Monodon monoceros*). Greenlanders have a long history of taking small cetaceans for their livelihoods.[10] Catches of beluga and narwhal, in particular, are important to many Greenlandic communities, both as food and as a source of cash derived from the sale of meat, *mattak*, and the narwhal's distinctive tusk. Regulation of small cetacean catches in Greenland is based upon a combination of Inuit customary law and municipal and Home Rule regulation.[11]

The fact that small cetaceans are not subject to IWC jurisdiction is one of the most contentious issues facing that body today. Species like beluga and narwhal were not included in the ICRW Schedule's annex of covered species when the convention was negotiated; however, since the adoption of the commercial whaling moratorium in 1982, debate has frequently focused on whether small cetaceans should come under international scrutiny. Denmark and Greenland argue that small cetaceans are a domestic matter because they live primarily within Greenland's two hundred-mile exclusive economic zone (EEZ).

This debate took on greater significance when biologists became increasingly concerned that current catch levels of beluga in waters between Greenland and Canada may be above sustainable limits.[12] In 1989, Greenland and Canada signed a bilateral memorandum of agreement about research, management, and use for these stocks. This agreement created a Joint Greenland-Canada Commission for Conservation and Management of Beluga and Narwhal, which recommended that Greenlandic hunters reduce their overall beluga catch and restrict the narwhal catch to current levels.[13] The Home Rule government revised its regulations in 1995 to respond to these recommendations.

The issue of small cetaceans amplifies ideological debates in the IWC about the ethics of marine mammal hunting generally. Many Greenlanders believe that the IWC's "Like-minded Group" is dominated by those with animal rights perspectives. In recent years, whale protection groups have argued that there are powerful ethical reasons not to whale, no matter whether the take is sustainable or not. A spokesperson for The Whale and Dolphin Society states that:

The use of whales for food is something which is not warranted because of the fact that whales are animals with special characteristics. They are a much more important source of inspiration to humanity; they are a source of beauty to humanity; they are sentient animals aware of their lives in the sea. Therefore they are animals which are fit subjects for moral concern. It is on moral grounds that The Whale and Dolphin Society opposes whaling.[14]

The US Marine Mammal Commission has also given credence to these arguments by suggesting that ethical issues should perhaps take precedence over scientific findings in the IWC.[15]

Much of this debate focuses on whether or not whales are—as one writer argues—"uniquely special" members of the animal kingdom.[16] If they are, then where is the boundary between them and those animals that are not? Does this uniqueness apply to both large whales and to smaller cetaceans (e.g., beluga or harbor porpoises)? Could it also apply to other marine mammals, such as seals or walrus? And if whales are unique, do they then—as many argue—have an individual "right to life"? Scientists debate what criteria could be used for making such determinations, and there is considerable disagreement about these issues. Sadly, the debate surrounding them is becoming ever more strident, suggesting that there is little room for differing views. Like the abortion controversy in the United States, the argument that an individual whale has a "right to life" effectively forecloses all thoughtful discussion. These ideological conflicts are beyond biology; rather they focus on ethics and morals, or more accurately, whose ethics and morals will prevail. Perhaps it is not surprising that these arguments come at a time when the IWC's new procedures and improved data may open possibilities for resumed whaling on certain stocks on a sustainable basis.

Greenlanders find these Euro-American attitudes about marine mammals to be hypocritical because they are applied selectively and inconsistently. In many parts of the industrialized world, whales have become a symbol for environmental concern. They, along with other so-called "charismatic megafauna" (harp seals, dolphins, pandas, and elephants), are photogenic and have enormous popular appeal.[17] Some face serious conservation challenges, whereas others do not; but these species have become the focus of campaigns that go beyond questions of conservation and sustainability.

Greenlandic author Finn Lynge questions how "universal" these attitudes are and points to troubling parallels between opposition to Greenlandic whaling and seemingly discredited colonial attitudes of the past. In his view, it is not surprising that some of the strongest anti-whaling and anti-hunting voices come from the United States, the United Kingdom, Australia, and New Zealand:

[W]e all know that the Anglo-Saxon countries have a strong tradition for wanting to export their own cultural pattern and value systems to everybody else, so it comes as no surprise that an essentially western urban phenomenon is presented as "all but universal." It does give rise though, to some questions: if cultures do exist in which respect for an animal and the killing of it go together, are they not entitled to go on existing? If not, what is the ethical reason for liquidating these cultures?[18]

This argument might seem overly harsh until one considers that there is a strong undercurrent of "social-Darwinism" in the rhetoric of some opposed to whaling. Roger Payne of The Whale and Dolphin Society states that "you can tell quite a lot about a country by how it treats other animals and that the greater the treatment—the more concern that countries give to animals—the perhaps farther along they are in the development of their own ideas."[19] That this comment comes from someone living in a country whose standard of living demands unprecedented and, in all likelihood, unsustainable "mining" of the earth's resources is not lost on Greenlanders. It leaves them wondering how anyone so situated could infer that his society's ideas about human-environment relationships are somehow "further developed."

Greenlanders are aware of humankind's need to kill for food. Living in a harsh and unforgiving environment, the realities of life and death are never far away. The close connection between food and the spiritual world is well understood; it is reflected in the words of an Inuk elder who reminded Knud Rasmussen that "the greatest peril of life lies in the fact that human food consists entirely of souls."[20] The fact that we kill for food no matter where we live is difficult for many Euro-Americans to understand, particularly those living in a highly urbanized environment. Even vegetarians who live solely on grains and fruits contribute to habitat destruction and the environmental contamination associated with agriculture.

Given Greenlanders' experience with these controversies at the international level, Home Rule resistance to IWC management of small cetaceans comes as no surprise. Having to deal with the IWC on small cetaceans would only exacerbate frustrations that Greenlanders already feel. Said Ingmar Egede, a Greenlandic participant at a recent IWC meeting:

As a Greenlander you sit there [in IWC meetings] with a bitter taste in your mouth: We got what we wanted, but [we] are still angry. We got a small rise in our quotas of minke and fin whales, on reasonable conditions. Maybe some years ahead we can even drag home bigger quotas. . . . [But] the uncertainties about the stock assessment put on these limits, and maybe the majority in the commission want to remind us about our dependency on their goodwill.[21]

This comment highlights the inequities in political power that underlie much of the debate about animal rights and humane killing. It is one thing for a country to decide on ethical or moral grounds that it will not catch whales, but is it appropriate to impose that view on societies that have a different perspective? In the IWC context, powerful nations use the threat of trade restrictions as a weapon to impose their views on others. The United States, in particular, has threatened to use economic sanctions against other nations based on provisions in the Marine Mammal Protection Act (MMPA) and the Pelly Amendment.[22] Indeed, in 1993, the US secretary of commerce certified that Norway's decision to resume commercial takes of minke whales despite the IWC's moratorium had "diminished the effectiveness of the International Whaling Commission (IWC)." He did so even though Norway had earlier lodged an objection to the moratorium and hence, under the IWC rules, was not bound to adhere to the moratorium decision. President Clinton responded to this certification by asking that a list of economic sanctions against Norway be developed. He stopped short of imposing sanctions, however, noting that "I believe our objectives can best be achieved by delaying the implementation of sanctions until we have exhausted all good faith efforts to persuade Norway to follow agreed conservation measures." His letter to the Congress goes on to state unequivocally that the United States is opposed to all commercial whaling, but he notes that the United States

has an equally strong commitment to science-based international solutions to global conservation problems. The United States recognizes that not every country agrees with [the United States] position against commercial whaling. The issue at hand is the absence of a credible, agreed management and monitoring regime that would ensure that commercial whaling is kept within a science-based limit.[23]

The threat of economic sanctions has been a significant factor in obtaining compliance with the IWC rules, for both commercial and aboriginal subsistence whaling. For Greenland and other countries dependent on fisheries exports, imposition of these sanctions could have a devastating impact. Greenlanders are not oblivious to this fact. As Greenland's minister of fisheries and industry states:

> I think that leaving the IWC could harm our overseas trade. If we continued whaling after opting out, it would be considered criminal by other countries. It would harm our exports and decrease understanding of our need for whaling and whale meat. So at present, we have no plans to leave the IWC.[24]

There are signs that the ideological polarization between Greenlandic hunters and some environmental organizations may be easing. For example, Premier Lars Emil Johansen met recently with Greenpeace representatives to explore areas of mutual concern. The meeting took place after Greenpeace publicly expressed regret about damage caused to Greenland by the anti-sealing campaigns of the 1970s and 80s. After the meeting, Johansen stated:

> Greenpeace isn't an animal rights organization, as many people seem to believe. It is an environmental organization, and Greenland and Greenpeace have similar interests in several areas. . . . If our environmental people and Greenpeace could work together on solutions to these problems, it could be very exciting.[25]

The Greenpeace representative at the meeting stated later that "We have never had anything against the Greenlandic seal hunt. This is a whipped-up conflict [with little basis]. There isn't anything that separates [Greenland and Greenpeace] on this issue."[26] Not all Greenlandic hunters likely share Premier Johansen's optimistic view about collaboration with environmental organizations. His views, nevertheless, are based on the idea that hunters and others who live close to nature *should* be able to form alliances with groups that care deeply about the environment. Greenlanders and environmentalists share common concerns in many areas: Arctic environmental contamination, nuclear proliferation, global climate change, and maintenance of biodiversity and protected areas. Certainly these issues could be the focus of common endeavors. At the very least, political realities in the world today suggest that a continuing dialogue between Greenland and environmental groups is useful to help overcome misunderstanding and to explore areas of mutual concern.

Conflict in the realm:
Greenlandic-Danish disputes about whaling

The swirl of politics associated with whaling is not limited to Greenland alone but also has profound implications for relations within the Danish realm as a whole. In recent years, conflicts over whaling policies have caused something of a crisis in Danish political circles. This conflict came to a head both in 1991 and again in 1994, when powerful Danish forces opposed to whaling came into direct conflict with Greenland and the Faroe Islands. At issue in 1991 was whether the Danish IWC delegation, including Greenlandic and Faroese representatives, would have the option of supporting limited commercial whaling in votes taken at the 1991 IWC meeting in Iceland. After the IWC's moratorium went into effect, Norway, Iceland, and Japan pressed for resumption of what is referred to as "small-type whaling" (STW). In the North Atlantic, this hunt is largely focused on minke whales and is carried out by smaller coastal vessels not unlike those used in Greenlandic whaling. The IWC, however, repeatedly rejected all such proposals, arguing that there was insufficient biological data about the affected stocks.

At the 1990 IWC meeting, however, the Scientific Committee acknowledged for the first time that whaling could be justified on biological grounds for at least one North Atlantic minke stock—those in the East Greenland–Iceland–Jan Mayen region. This finding was a sign of a significant turning point in IWC deliberations, since it became clear that objections to whaling in the future might not be sustained solely on biological grounds.

In the spring of 1991, both Greenland and the Faroe Islands pressed the Danish Foreign Ministry to issue instructions to its IWC delegation to support limited STW if the matter came to a vote in the IWC plenum. These instructions are normally issued after negotiations between Denmark's ministries of foreign affairs, environment, and fisheries, and Home Rule officials in both Greenland and the Faroe Islands. After a weeklong meeting in May of 1991, the Foreign Ministry agreed with the Greenlandic proposal and issued instructions that would enable Denmark to support STW if it could be conducted sustainably.[27] Explaining why Greenland, which conducts aboriginal whaling, should press for a controversial resumption of commercial whaling, Ingmar Egede (who represented the ICC) stated:

We [Greenlanders], coming from a hunting tradition, understand the arguments and the frustrations of the Icelanders, the Norwegians and the Japanese. Their calculations [of affected whale stocks] look honest to a layman, and we have an immediate understanding of and sympathy for their argumentation, and we realize that we are left back with the same anger and frustration as they express.[28]

When anti-whaling activists in Denmark got wind of the Foreign Ministry's instructions, the reaction was swift. Said Margrethe Auken, a former member of the *Folketing* and an outspoken opponent of commercial whaling,

if there isn't an immediate outcry from [environmental] organizations, politicians and others, . . . [Danish officials] together with the Greenland and Faroese home rule governments, will calmly and quietly ease us back into the "whalers club" . . . which we, with much effort, managed to get Denmark out of 10 years ago.[29]

Michael Gylling-Nielsen of Greenpeace Denmark commented that Greenland's support for the measure could have dire consequences:

We will not campaign against Greenlandic whaling. That, I want to emphasize. But if Greenland supports commercial whaling, then we cannot close our eyes. . . . Greenland is doing itself a disservice when it supports [whaling by] Iceland, Norway and Japan. Greenland ought to ally itself with the majority [of "Like-minded Nations"] in the IWC against wishes or demands about commercial whaling, because Greenland is dependent upon the majority for its own [aboriginal whaling] quotas.[30]

Under a banner headline reading "Greenland is threatened," a Greenlandic newspaper editorial responded sharply to this criticism, pointing out that this submerged threat was not limited to Greenland's whaling quotas alone:

[Greenpeace's] next goal is to stop our fisheries. If Greenland doesn't join efforts against these organizations' single-minded perception of reality, and draw attention to itself as a country that also has a right to exist, then one day we'll wake up like Sleeping Beauty to find that it is too late.[31]

Just before the 1991 IWC meeting, the Danish Parliament took up the question of the IWC instructions. In a tense exchange, the Social Democratic majority opposed both small-type coastal whaling and the Foreign Ministry's instructions.[32] Although they supported Greenland's request for additional aboriginal quotas, they opposed STW as "contrary to North Atlantic interests." Reactions from Greenland's political leadership were fierce. Greenland's prime minister blasted the majority's actions:

this borders on imperialism, which we thought had disappeared. It is shocking that our closest partners [in the *Folketing*, i.e., the Social Democrats] who use Greenland's interests in their own internal power struggle. This is a serious matter, which can easily have consequences for our working relationships.[33]

Hans Pavia Rosing, one of Greenland's two *Folketing* members, quoted author Finn Lynge's observations about Greenland's relationship to the realm:

Denmark has not demonstrated that it has any special authority over the Faroes or Greenland. The Danish realm consists of three very different people in terms of ethnicity, language and culture, and colonial attitudes from earlier times don't mean anything. Greenland and the Faroes can no longer be put in their place like the old days, and certainly not when it comes to the old hunting traditions.[34]

Denmark's foreign minister himself professed deep concern about the crisis:

I understand the bitterness Greenlanders feel. We need to watch out that we don't destroy the realm. We have heard from both the Faroe Islands and Greenland that they believe there are some things in the realm that we must reconsider. . . . I don't believe people down here [in Denmark] really understand what whaling means for these people.[35]

In the end, the IWC voted to continue the moratorium on commercial whaling, thereby scuttling small-type whaling for the time being. The vote was eighteen members for continuation of the moratorium and six against. At the last minute, the Danish foreign minister instructed the delegation to abstain but emphasized that the government's position had not changed. "Had it been possible to get a majority now, and had Denmark's vote been crucial," he said, "then Denmark would have voted for [small-type whaling]."[36]

A similar conflict erupted again in 1994, this time focused on the issue of an Antarctic whaling sanctuary. Whale preservation groups had strong support for a complete ban on commercial whaling in the Antarctic, even though the Scientific Committee had not been allowed to comment on the idea. As negotiations surrounding the proposal moved ahead in the spring of 1994, it became clear that Denmark's vote on the proposal could be crucial to the outcome. After hectic negotiations leading up to the IWC meeting, Greenland's premier gave conditional support to the Antarctic sanctuary, a decision that generated considerable controversy in Greenland itself. Premier Johansen gave his support after gaining Danish agreement on three points:

1. that the Danish delegation would work to ensure that establishment of an Antarctic sanctuary would not serve as a precedent for establishing a similar sanctuary in the Arctic;
2. that if Denmark's vote became crucial for approval of the Antarctic sanctuary, then Denmark would work for a package solution that would end the moratorium so that commercial whaling could take place on a sustainable basis; and
3. that Denmark would work for an increase of Greenland's minke whale quota.[37]

The Home Rule leadership's decision to give qualified support to the sanctuary drew sharp criticism from KNAPK, Greenland's organization of hunters and fishers. KNAPK's president stated that the decision undermined Greenland's opportunities for raising the minke quota and thwarted efforts for greater cooperation between Nordic countries interested in sustainable whaling. KNAPK's leadership may also have had in mind the potential interest of some in Greenland in exporting whale products, it is could be shown to be sustainable.[38] In a sharp rebuttal, Premier Johansen argued:

It is important that we not just lock ourselves in and shout and isolate ourselves from the rest of the world. We have to think both strategically and tactically, and by supporting the whale sanctuary Greenland will get significant international goodwill which can be beneficial to our own whaling.[39]

As it turned out, Greenland's and Denmark's support for the sanctuary proposal succeeded in bringing at least some of the goodwill they hoped for. At the 1994 meeting, the IWC approved an increase in Greenland's minke whale quota over a three-year period, effectively increasing the possible take from 115 per year to 155.

Greenland's relationship with Denmark over whaling issues could well become even more complicated in the years ahead because of the latter's membership in the European Union. For example, some have suggested that the EU take a common position on whaling issues in the IWC. Greenland's hunters express strong concern about this proposal, especially since some EU countries are vocal opponents of whaling:

There is no doubt that a unified EU voice and policy on the IWC would have an adverse impact on issues relating to Greenland's whaling interests. Therefore, we have a reason to fear that Denmark as a member of the EU may take positions contrary to the interests of Greenland.[40]

These controversies illustrate the sympathy that most Greenlanders feel for other nations interested in resuming whaling, but they also demonstrate the price that Greenland could pay for supporting them. The concerns clearly show Greenland's political marginalization with regard to whaling within the Danish realm and beyond. As one politically active Greenlander notes (speaking anonymously):

Whaling is the only exception within the area of living resource management where the Danish realm has not worked out a more autonomous relationship for Greenland. Greenland simply must sit in the background at IWC meetings. . . . The decision that whaling policy should be determined by all entities in the realm should have been discussed more thoroughly in Greenland . . . because Denmark and Greenland have different histories and backgrounds with regard to whaling.

Recognizing this marginalization, Greenlanders are actively developing other strategies that they hope in the end may prove more productive.

Greenland and the North Atlantic Marine Mammal Commission (NAMMCO)

The conflicts described above contributed to Greenland's decision in 1992 to join other nations in creating NAMMCO, the North Atlantic Marine Mammal Commission. Greenland's fisheries minister joined counterparts in Iceland, Norway, and the Faroe Islands to create this new entity on April 9, 1992, "to contribute through regional consultation and cooperation to the conservation, rational management and study of marine mammals in the North Atlantic."[41] NAMMCO operates at the level of fisheries ministers in signatory countries rather than at the government level. This setup was intentional because of Greenland and the Faroe Islands, which remain part of the Danish realm. The organization consists of a council, management committees, a scientific committee, and a secretariat based in Tromsø, Norway.

NAMMCO grew out of a series of meetings and conferences on management and rational utilization of marine mammals held between 1986 and 1991.[42] Initially, several of the parties involved considered creating an alternative to the IWC; however, NAMMCO's objectives

were ultimately scaled back to a less ambitious agenda of regional cooperation on management of marine mammals not subject to IWC jurisdiction. Several participating parties emphasized that they continue to
view the IWC as the appropriate management entity for larger whales.
Whale preservation groups were nevertheless highly suspicious of
NAMMCO and pressured Greenland to stay out of it.[43] This tactic
ended up having the opposite effect, however. Greenland expanded its
involvement, citing longstanding dissatisfaction with the IWC's structure. Said KNAPK's chair:

It is difficult for us to accept that those decisions are taken by a majority of
nations who have no knowledge of and no primary interests in whaling and
marine mammal hunting, and often have no national waters at all. For that
reason, the IWC should be revised to consist of a forum of nations utilizing
marine mammals—and perhaps just representing Northern marine resources.[44]

NAMMCO held its first meeting in the Faroe Islands in September
of 1992. At that meeting, signatories approved the organization's basic
structure and established a secretariat. The council consists of one representative from each contracting party, and its purpose is to "provide
a forum for the study, analysis and exchange of information" about marine mammals.[45] Council actions require unanimous approval from
those present. The composition of the scientific committee was also
approved: There are twelve members, with three appointed from each
contracting party. The committee's initial tasks included assessing the
status of selected marine mammal stocks in the region, including pilot
whales, northern bottlenose whales, harp and hooded seals, killer
whales, and Atlantic walruses. In subsequent meetings, discussions focused on NAMMCO's relationship to other entities, particularly the
International Council for the Exploration of the Sea (ICES).
NAMMCO's council decided to cooperate with ICES at the working
group level and to arrange reciprocal observer status with the organization. The council also discussed the role of management committees,
which are to define needs for scientific assessments and to formulate
management strategies to implement council decisions. At its most recent meeting, NAMMCO also established working groups on hunting
methods and on catch inspection and observation.

According to Hoel, NAMMCO was created in response to at least
five issues.[46] First, the contracting parties were increasingly concerned
about the influence of whale preservation interests in the IWC, re-

flected in the IWC's initial reluctance to approve a new RMP. Second, NAMMCO signatories, coming as they do from fishing nations, wanted to strengthen multispecies approaches in management rather than simply focus on one category of animal or one species. Third, the parties recognized the need for greater management of small cetaceans but found the IWC context wanting in this regard. This issue is linked to the fourth concern, which is what Hoel calls "creeping jurisdiction" of the IWC, particularly toward management of small cetaceans. A final issue is the need for improved information and communication about marine mammals in the North Atlantic.

Clearly, NAMMCO was established because of strong dissatisfaction with the IWC regime. Because signatories viewed the IWC as moving away from science-based decisionmaking, they sought to establish their new organization in accordance with the principles of international law articulated in the UN's Law of the Sea Convention (discussed in chapter 4). UNCLOS entered into force in 1994 and is regarded by many as the most comprehensive international law ever enacted.[47] Chief among its provisions is the two hundred-mile Exclusive Economic Zone (EEZ) and also clear recognition that coastal nations have "sovereign rights for the purpose of . . . exploiting . . . the natural resources, whether living or non-living" within their EEZ. As noted in chapter 4, governments involved in the UNCLOS negotiations resisted applying the concept of "common heritage of mankind" to marine animals on the high seas. They also resisted establishing a single global oceans management body with a strong international secretariat. Instead, UNCLOS's Article 65 states that nations are to "cooperate with a view to the conservation of marine mammals and in the case of cetaceans shall in particular work through the appropriate international organizations for their conservation, management and study." This reference to "organizations" (plural) creates ambiguity about which international entities (beyond just the IWC) might have jurisdiction over marine mammals. NAMMCO signatories believe that this ambiguity leaves ample room for multiple international organizations, particularly at the regional level.

These conclusions are challenged by NAMMCO's detractors, including Sidney Holt, an outspoken opponent of commercial whaling. He argues that NAMMCO's bylaws conflict with UNCLOS because membership is restricted only to nations approved by the original founding members: "NAMMCO . . . will admit only new member countries by agreement of the existing Members, effectively to bar the participation of the other range states—the majority, of course—which do not now engage in whaling."[48] He further argues that, if NAMMCO expands its mandate, it will contradict agreements reached at the 1992 UN Conference on Environment and Development (UNCED) in Rio. This conference, he argues, "recognized, by consensus, the IWC as the sole international body having such competence." NAMMCO parties, however, found no such consensus at Rio and reject Holt's argument.[49] They point to provisions of *Agenda 21* (the UN's environmental action plan approved at Rio) as reinforcing their organization's goals. Specifically, they note language in *Agenda 21*'s chapter on marine affairs, which states that the Law of the Sea Convention "provides the international basis upon which to pursue the protection and sustainable development of the marine and coastal environment and its resources."[50] It also affirms that "states commit themselves to the conservation and sustainable use of marine living resources" and goes on to recognize the responsibility of the IWC and other international organizations for the conservation, management, and study of cetaceans.[51] In Hoel's view,

There can be no doubt that NAMMCO has a firm basis in *Agenda 21*, not least by contributing to fulfill the duty that "States should cooperate for the conservation, management and study of cetaceans" (Articles 17.63/17.91), in that stocks not currently managed come under a management regime.[52]

Hoel sees the possibility of productive relationships developing between the IWC and regional management bodies like NAMMCO, but of a different type than those now in place:

The major problem in international whale management is now, as it was ten years ago, that the real threat to marine mammals is not harvest, but the impacts from pollution, bycatch in fisheries, seismic survey shooting, nuclear test explosions, and the like. A relevant future policy area for the IWC is therefore the formidable task of informing its member governments of the effects of such activities on whales, while the management issues could be shifted to appropriate regional organizations that can manage whales on a sustainable basis in relation to their role it the ecosystem.[53]

Greenland's decision to join NAMMCO presents both risks and opportunities. By aligning itself so closely with nations interested in commercial whaling, Greenland risks confronting the "Like-minded Nations" and anti-whaling groups. Greenland's strategy of joining NAMMCO while remaining active in the IWC can be viewed as a form of "cautious resistance and calculated conformity" like that described by Scott in *Weapons of the Weak*. By being involved in both organizations, Greenland appears to be protecting its economic interests while offering political support to those dissatisfied with the IWC. This calculated move is made possible by the existence of Home Rule and demonstrates the opportunities open to Greenland because of self-determination. Without Home Rule, Greenlanders would be forced to work solely through Danish government channels on all international matters. Because of its autonomy, Greenland is able to create alliances with other marginalized nations and organizations on matters of mutual interest. Paraphrasing Feit (cited in the introduction), Home Rule enables Greenlanders to mobilize political and economic leverage in arenas beyond its borders or resist unfair external pressure and to restructure relationships with distant political and economic institutions.[54]

Multiple strategies: Greenland and the Inuit Circumpolar Conference Whaling Agenda

Greenland is not limiting its whaling initiatives to NAMMCO alone. While pursuing North Atlantic alliances, the Home Rule government is also actively involved with the Inuit Circumpolar Conference (ICC) on whaling issues. The ICC, which was established in 1977, represents Inuit people living in Greenland, Canada, Alaska, and Chukotka.[55] Whaling issues have always been at the forefront of its agenda, originating in conflicts over the Alaska Eskimo bowhead controversy in the late 1970s. A resolution from the ICC's first meeting in 1977 reads:

Whereas, the Inuit have hunted the whale for thousands of years, and the relationship between the Inuit and the whale has become a necessary part of the Arctic ecological system; and
Whereas, there are those who do not understand the relationship between the Inuit and the whale, and are working to stop Inuit whaling as a means of preserving whale species being destroyed by commercial whaling; and
Whereas, Inuit whaling is subsistence whaling and not commercial whaling; and
Whereas, whaling is a necessary part of Inuit cultural identity and social organization, and is in no way similar to commercial whaling;
Now, therefore, be it resolved that the delegates assembled at the first Inuit Circumpolar Conference call upon the United States and Canadian delegates . . . to defend Inuits' aboriginal right to hunt the whale in the Arctic.[56]

The ICC has had observer status at the IWC since 1980. Early observers emphasized the need for a more aggressive ICC strategy with regard to whaling issues, even raising the idea of an Inuit circumpolar whaling commission. This idea, however, was not universally supported within the Inuit community, and the ICC's leadership focused more on issues of sustainable use of Inuit resources and lands.

The issue came up once again in the ICC's 1992 General Assembly, and delegates adopted a resolution requesting that

. . . the ICC General Assembly take immediate steps to initiate the operation of the ICC Whaling Commission by instructing the ICC Executive Council to activate this commission and to determine its composition, structure, and to develop terms of reference [and]

. . . the ICC Executive Council . . . promote discussions among interested parties within the circumpolar world including a consideration of international cooperation with regional organizations such as [the] Canada/Greenland Joint Commission [on Narwhal and Beluga] and the North Atlantic Marine Mammal Commission.[57]

Les Carpenter, at the time ICC vice president from Canada, stated that Inuit peoples, including Greenlanders, ought to form their own whaling commission: "Personally, I think that we ought to form our own whaling organization, because the anti-whaling nations in the IWC will eventually go after aboriginal subsistence whaling."[58]

At the 1995 ICC General Assembly in Alaska, the ICC's leadership followed up on the earlier resolutions with a document entitled "Circumpolar Whaling and the ICC Whaling Agenda."[59] In it, then vice president Ingmar Egede expressed concerns about the political climate in the IWC with regard to Inuit whaling:

As Western societies over-did their whaling in the past, so they also over-do the protection of whale stocks today. They interfere in the numbers of whales we take, they call our hunting methods inhumane. For some reason, they do not trust our capability to utilize one of our most important food resources in a sustainable manner.[60]

The document goes on to note that

the problems that [the] IWC experiences in discharging management responsibilities in accordance with international customary law and accepted regulatory practices raise questions about its usefulness and moral legitimacy. . . . The willingness of the majority of IWC members to work openly against the purpose of the international whaling convention . . . suggests that the status of the IWC as an "appropriate" management organization can, with abundant justification, be seriously questioned.[61]

In the 1995 General Assembly, delegates considered the options available to them for alternative whaling management regimes. The ICC

Whaling Agenda document asks: "Are the present and continuing rights, needs, and interests of Inuit whaling communities adequately protected, or can they be better protected by establishing (or by helping establish) a new international whaling authority?"[62] The document suggests that any such authority would have to

- embrace sustainable and equitable resource use goals
- respect diverse cultural traditions and national sovereignty
- incorporate traditional ecological knowledge into co-management arrangements

The document also emphasizes the importance of new international agreements supporting the sustainable use of living resources, including whales. *Agenda 21*, in particular, urges governments to (1) cooperate with, or establish subregional and regional management organizations, (2) encourage sustainable use of marine resources by local communities and aboriginal peoples for food and as sources of income, (3) incorporate traditional knowledge into management and legislation, and (4) recognize indigenous rights to subsistence. The 1991 World Conservation Strategy, endorsed by the World Conservation Union (IUCN), the UN Environment Program (UNEP), and the World Wide Fund for Nature (WWF), also stresses the importance of regional cooperation in ensuring that local people are able to use and manage wild resources sustainably.

In 1991, the eight Arctic nations also signed the Arctic Environmental Protection Strategy (AEPS), which establishes principles for Arctic cooperation. As part of the AEPS process, member nations signed the 1993 "Nuuk Declaration on Environment and Development in the Arctic." This document affirmed that *Agenda 21* should be the basis for Arctic cooperation and that (1) states have sovereign right under international law to exploit their natural resources, and (2) indigenous peoples have a vital role in environmental management because of their traditional ecological knowledge. The newly-formed Arctic Council, formed in September 1996 by the eight nations, is expected to abide by these same principles.

At the 1995 ICC General Assembly, delegates put off making any major decisions about whaling policy. For Greenland, however, political considerations must certainly be a major factor in how it explores alternatives to the IWC. As a nation largely dependent on fisheries exports, Greenland could well encounter serious challenges from antiwhaling groups if it left the IWC. As Greenlander Ingmar Egede states: "We are so dependent on the powerful nations [in the IWC] that we can't leave. So long as that's the situation, they can strangle us economically. And therefore, we have to think carefully before the ICC can recommend any changes."[63]

It remains to be seen what might come from this ICC debate about whaling policy, but Greenland will undoubtedly continue to be a major player in these discussions. These two Home Rule initiatives—the creation of NAMMCO and support for greater ICC involvement in whaling policy—illustrate the importance of Home Rule and self-determination in protecting indigenous whaling. The difficulties within the Danish realm over whaling make clear that, without Home Rule, Greenlanders would be even more vulnerable on whaling issues. As a part of the Danish realm, Greenland has the advantage of using Danish political institutions and processes when it is useful to do so, but it is not entirely restricted to them. Home Rule also enables Greenlanders to work through international fora to create alliances with other, similarly marginalized nations and groups.

Conclusion: Whaling and Sustainability in Greenland

In the aftermath of the 1992 UN Conference on Environment and Development (UNCED) in Rio, the world's attention has been focused increasingly on promoting development that is both sustainable and equitable. The Brundtland Commission's now well-known definition of sustainable development—that which meets the "needs of the present without compromising the ability of future generations to meet their own needs"—is widely recognized as an important starting point for debate.[1] Although the concept of sustainability is somewhat problematic, it nevertheless focuses our attention on the importance of maintaining healthy ecological systems and meeting the essential needs of human societies.[2] It also implicitly acknowledges the social nature of human-environment relations and the historical dimension of resource conflicts.

In the introduction, I made several major arguments about the relationships between Euro-American and Greenlandic societies as they affect whaling. First, I argued that historically informed theory and theoretically informed history are vital to understanding relationships between sustainability and indigenous self-determination. Greenland's history shows how different modes of production—what Eric Wolf characterizes as "the political-economic relationships that underlie, orient, and constrain interaction"—give rise to divergent systems of ideas.[3] The mode of production characteristic of contemporary Greenlandic communities (described in chapter 2) continues to be largely kin-oriented. The mixed economy in Greenlandic communities is one where cash and country food production complement each other, even when wage employment and transfer payments are significant. Cash is used in this context not so much for amassing huge wealth but rather for promoting the mutual security of families and extended families across generations. Tendencies toward greater social differentiation are mediated by culturally based leveling mechanisms involving sharing and exchange.

Greenland's history and mode of production inform a system of ideas about the appropriate use of living resources, including whales. Greenlanders desire to eat locally available, nutritious, and culturally valued foods, just as their ancestors have done for generations. Doing so means killing whales and other marine mammals. With this activity comes an awareness of the responsibility they have toward conserving these resources and sharing them with others. Generations of experience in the Arctic have produced the realization that ecological conditions can change almost overnight, and that flexibility and adaptability are essential ingredients for survival. Flexibility means holding to the belief that new technologies may well be necessary to cope with changing conditions and to take advantage of new opportunities. Not every hunter in Greenland can perhaps articulate these ideas, nor are they held consistently by all; but by and large, these ideas and practices have enabled Greenlandic society to sustain itself for generations.

As chapters 4 and 5 make clear, these ideas and practices often conflict with those of Euro-Americans. I argue that policy conflicts within the IWC about the ethics of whaling, the morality of exchanging whale products for cash, the use of new technologies in whaling, and the incorporation of indigenous knowledge in management are informed by these conflicting idea-systems. For many Euro-Americans, whales are perhaps the most important symbols of environmental concern. "Saving the whale" epitomizes what must be done to stop global environmental degradation. To imagine killing a whale is to raise fears that industrial societies will slide too easily back into a pattern of ruthless exploitation of open access resources. This thought is especially disconcerting to those who live in urban environments, far removed from settings where whales are frequently seen and human societies still rely directly on the sea and the land for sustenance. For hunters in Greenland, however, whales are often as common in daily life as songbirds are in an urban backyard. Like all forms of life, they are worthy of respect and concern, but one also grows up around them, sees them when out in a boat, and eats them as a matter of course. In this context, taking the life of a whale to feed one's family is an affirmation of the interdependence of all life in a demanding Arctic environment. It is done with an understanding that life and death are closely linked.

Likewise, many Euro-Americans view exchanges involving cash as being devoid of social or cultural dimensions. "Cold hard cash" buys gas for the car, food at a supermarket, or an airline ticket for a vacation. Increasingly, "cash" in Euro-American societies is simply the swipe of a credit card through an automated teller machine. In these circumstances, exchanging cash for goods and services demands little or no relationship between the buyer and the seller. It is thus no surprise that the presence of cash in exchanges involving whale meat or *mattak* in Greenland generates suspicion among some that "profit" is the sole motive in these transactions. If cash is involved, are not these transactions merely "commercial"? For Greenlanders, though, cash is simply one of several mediums for creating and maintaining relationships with family and extended family members, with fellow Greenlanders, and with the wider world. Of course, cash is used to buy Mariner outboards and Sako rifles from distant suppliers, but cash also enables local societies to reproduce themselves. People with wage jobs in a small community purchase whale meat from hunters, who use that cash to buy gas for outboards, shells for rifles, winter boots at the store, and perhaps a video for their children. This interdependent relationship provides hunters with the means for procuring resources and consumers with culturally valued foods. Like gift-giving and sharing within families, it reinforces family and community solidarity, extending even to Greenlanders in distant towns and settlements.

Similarly, many Euro-Americans (particularly tourists) express disappointment when they find that many Greenlandic hunters no longer use kayaks or hand-thrown harpoons in catching whales. The image of "Eskimos" living on the edge of existence is seared deeply into our collective subconscious; we want to know that there are still people in the world "living in harmony with Nature." Just as a farmer is still a farmer, however, whether driving a John Deere tractor or a plow behind a team of mules, so are hunters still hunters, even if they use more sophisticated technology. I do not mean to say that technology does not change hunting practices; I simply want to make the point that these changes alone do not necessarily lead to fundamental changes in livelihood. Greenlanders today have adopted many new forms of whaling technology, just as their ancestors did when Thule peoples arrived nearly one thousand years ago. Flexibility and adaptability are hallmarks of Inuit cultures, and the adoption of new technologies by hunters reflects a pragmatic approach to life in a demanding environment.

Many resource managers are skeptical about the value of hunters' knowledge in managing whales and other living resources. Although a lot of lip service in the North is given to incorporating indigenous knowledge into resource management, the reality is that there are relatively few examples where this synthesis has taken place successfully. In an age of tracking whales by satellite and mitochondrial DNA techniques for identifying discrete stocks, the mere "anecdotes" of hunters are, at best, difficult to reconcile with scientific data and, at worst, annoying distractions. Good hunters are good observers, however, and much of science is about careful and consistent observation. Good scientists understand the value of observations from those who are out on the ice or on the water daily. Often conflicts between university-trained resource managers and hunters arise because of a lack of respect or appreciation about the fact that each approach to learning has its strengths and limitations. Our limited experience suggests that these differing ideas and attitudes can be overcome when both groups recognize the value of sharing insights and observations stemming from very different experiences.

As these examples illustrate, there are striking differences in ideas, attitudes, and practices between Euro-American societies and Greenlandic society. These differences are to be expected, and they can be enriching for humankind as a whole. Problems arise, however, when we use the ideas, attitudes, and practices of one society to judge those of another, to assume that one society's ideas are "further advanced" than those of another. In the Greenlandic context, this mindset could deny Greenlanders a vital part of their own history.

A second major argument I make in the introduction is that Greenlanders' achievement of Home Rule was a significant step both toward self-determination and toward creation of effective co-management regimes. Home Rule enables Greenlanders to take the initiative in restructuring relationships with world political an economic systems. Recognizing this ability does not deny the power of internal colonialism or the persistent dependency of the periphery but is simply accepting, as Feit does, that the center is not "all-powerful and hyper-integrated."

The self-determination provided by Home Rule is a major contributor to an increasingly effective co-management regime for whaling. In the introduction, I presented Fikret Berkes's characterization of a well-functioning regime: one that is *efficient, stable, resilient,* and *equitable.*[4] Drawing upon the material presented here, I suggest that Greenland's whaling regime is becoming efficient in resolving disputes and maintaining compliance with quotas and regulations. Clearly difficulties remain, but the record from recent years shows a marked improvement in responsiveness at all levels. Undoubtedly, the political legitimacy enjoyed by the Home Rule state is a major factor here. It is difficult to imagine how effective conflict resolution over whaling could be accomplished without self-determination. The whaling regime has also shown itself to be stable; it has coped effectively with progressive changes in external and internal political demands, changing ecological factors, and the availability of new technologies. It has incorporated rather dramatic changes in quotas over the years and has implemented a major program for improving hunting technology. The regime has also demonstrated its resilience by accommodating a steady stream of surprises and shocks, from the occasional quota violation in recent years to challenges to its authority from some local hunters and politicians. The internal management system of the regime is, moreover, increasingly equitable, despite some disputes about quota allocations, especially between vessel and collective whalers. Internal conflicts have been addressed by giving more authority to municipal governments and by actively involving KNAPK and hunters themselves in policy debates and the regulatory process.

The real questions about the regime's overall equitability seem to lie at the national and international levels: To what extent will Greenlandic whaling be influenced by differing viewpoints within the Danish realm or within the European Union? Will Greenland have a more independent voice about whaling policy in the years ahead? And, to what extent will Greenlanders be able to pursue sustainable whaling within the context of the IWC? These questions remain unanswered, but clearly the future of Greenland's regime is linked to the successful resolution of these conflicts.

Although this case study shows that Greenland's whaling regime has become increasingly effective, it also reveals some of co-management's limitations. As I argue in the introduction, co-management cannot be viewed as a panacea for resource conflicts. It has significant transaction costs, not only in simple monetary terms but also in cultural disruption and stress. As the example from Qeqertarsuaq illustrates, hunters face a steady stream of changing quotas and regulations that cause stress and contribute to increased social differentiation. For an effective co-management regime to function, it must have credibility with local stakeholders—especially hunters themselves. As the ones with fingers on the trigger, they are arguably the "ultimate" resource managers. Demands on them that are perceived to be unreasonable or inequitable can lead to quiet resistance (e.g., violating quotas) or to active opposition. Similarly, co-management brings new pressures and demands upon indigenous governments. Attempting to respond to national and international expectations can create tensions that strain fledgling processes of self-determination and institution building. In a society of only 55,000 people, these tensions can be debilitating. Pointing out these difficulties is not to deny the strengths of co-management; it simply reminds us that these transaction costs must be taken into account when developing new regimes or evaluating those already in place.

It is thus important to recognize that co-management is a *process* and not an end in itself. Co-management is often used as a sort of buzzword in the Arctic today, perhaps leading some to think that simply including provisions in negotiated agreements may be sufficient for achieving sustainability. The road to co-management is not always smooth, however. We need to exercise patience in assessing the success or failure of co-management regimes. In the Greenlandic case, for example, it might be easy to criticize a single quota violation, but such episodes must be viewed from a broader perspective. As a process, co-management is inherently dynamic; tensions will exist, and mistakes will almost inevitably be made. The tensions involved may even be beneficial in the long run. As one Home Rule official in Nuuk acknowledged, "it isn't always nice to have people [in the IWC] interfering, [but] maybe it makes us do better what we would have done anyway."[5] Greenland's whaling regime has undergone dramatic transformations within the lifetimes of hunters alive today. It would be unfortunate to overlook the significant accommodations that they are making to new realities.

Whaling and sustainability:
Developing appropriate management regimes

The data presented here provide insights for advancing sustainability in contexts beyond just Greenland. Ecologists today are seeking a better understanding of human practices worldwide that protect biodiversity and promote healthy marine and terrestrial ecosystems. The same is true of social scientists—anthropologists, sociologists, and political scientists—who seek to understand what types of human-environment interactions promote sustainability and equitability. The intent of these complementary processes in the scientific community is to provide clues about how human societies can modify their activities to promote sustainable practices and to discourage those that are unsustainable.

In the context of the IWC, the adoption of the Revised Management Procedure (RMP) and the advances in scientific data about whales mean that debates will most likely focus increasingly on issues of ethics and morality rather than on biology. As William Aron, former US commissioner to the IWC, points out, the major conservation battles over whaling have been won:

> If concern is based largely on the need to correct abuses of the past and restore marine mammal populations to former abundance levels, continuing the current sweeping policy of virtually total protection for all species is no longer required. If, however, the concern for marine mammals is essentially based on the ethics or morality of their harvest, we probably should continue our current course. In doing so we must clearly recognize that there is a difference in imposing a moral or ethical standard on US citizens versus imposing such standards on the international community.[6]

If the "battle to save the whales" has largely been won, then what are the implications for those who continue to use whales? Will they be able to do so, and if so, under what conditions? Will whales increasingly be viewed as a part of a "global commons" and therefore subject to global regulation? Many find the global commons concept attractive because it focuses on concerns about global warming, ozone depletion, and marine contamination. At its best, it creates hope for finding collective solutions to pressing trans-boundary management problems. At its worst, however, the concept raises the specter of a global environmental management system based upon the notion of universally applicable knowledge, a system for global environmental enforcement, and a worldwide culture of global concern.[7] This system seemingly implies a universal culture where many, if not most, people share common global concerns over seemingly "irrational" local interests. In such a system, global knowledge may well take precedence over that which is "merely local."[8]

Greenlanders accept the importance of global thinking about the environment and development. Indeed, they and other Inuit are at the forefront of the Arctic Environmental Protection Strategy and were active players at the UN's Rio summit. They are also concerned, however, that the concept of "global concerns" not become simply a euphemism for "Euro-American concerns." Many believe this preemption is already happening with regard to whaling. Greenlanders cite as an example the rather extraordinary suggestion of animal rights activists

D'Amato and Chopra that they and other Inuit be paid large subsidies *not* to whale, and that

with these revenues, [Inuit] could migrate to places where food is more plentiful, or set up arctic farms, or—at least for a while—simply purchase food supplies. The . . . payments could be made over a limited period [because over
time] . . . the Inuit would be legally disabled from hunting whales sooner or
later, irrespective of the history of exploitative commercial whaling by other
nations.[9]

Greenlanders marvel at the ignorance of Arctic history, culture, and
economy reflected in these comments. They marvel, too, at the arrogance of those who would suggest that they simply pack up and leave
the Arctic in response to the qualms of those living to the south. Although D'Amato and Chopra's comments may be extreme, they illustrate the types of attitudes frequently encountered in the IWC context.
They also reinforce Greenlanders' concern about the equity of current
whaling regimes. Effective regimes presumably have a built-in system
of "checks and balances" enabling managers to respond to multiple demands in an equitable fashion, but these comments lead hunters to
wonder if, in the IWC context, there are too many "checks" on their
activities and too little "balance" in addressing their needs.

Research about resource management conflicts elsewhere in the
world suggests that an important step toward conflict resolution is to
recognize that all parties have legitimate points of view. Thompson and
Warburton, writing about environmental disputes in the Himalayas,
note:

Recognition of this pervasive heterogeneity immediately makes approaches
based on a homogeneous understanding of "the problem" inappropriate. . . .
In the single problem/single solution approach, the institutional reality at the
local level—the seething mass of contradictory problem definitions, con-
tending perceptions, divergent personal strategies and polarized policy pre-
scriptions—is inevitably part of the problem.[10]

Thompson and Warbuton argue that a problem's heterogeneity at
every level is "a rich resource, perhaps the ultimate resource."[11] The
solutions to these conflicts—like those surrounding whaling—lie in
recognizing that there are multiple problem definitions and multiple
solutions. Resolving these conflicts, they argue, "no longer requires us
to insist that one set of policy actors is right and that all others are
wrong. Instead, our attention is directed towards the notion of *appropri-
ateness*: which kinds of social transactions are best handled by which

institutional mode?"[12] As in the Greenlandic case, we have to ask which institutional modes can effectively encompass complex and contradictory social transactions driven by diverse histories, cultures, and modes of production. Goldsmith and others argue that the

key to [this] struggle is the building up of open and accountable institutions that restore authority to commons regimes—a struggle which requires increasing the bargaining power of those who are currently excluded or marginalized from the political process and eroding the power of those who are currently able to impose their will on others. Only in this way—when all those who will have to live with a decision have a voice in making that decision—can the checks and balances on power that are so critical to the workings of the commons be ensured.[13]

In this spirit, a group of social scientists (including myself) with a broad background in resource management conflicts recently examined whaling issues to explore which institutional modes and practices contribute to sustainable and equitable use and which do not.[14] The group did not focus on aboriginal subsistence whaling as such. Instead, their objective was to look beyond current categories of whaling (e.g., commercial, scientific, and aboriginal subsistence) used in the IWC to focus on which whaling practices might be considered "permissible" and "impermissible" when viewed from the perspective of sustainability. The rationale for this approach is that

the disintegration of the international whaling regime stems from the persistence of a method of distinguishing between permissible and impermissible whaling that is now regarded by many as arbitrary to the point of causing some member countries wishing to continue whaling to withdraw from the regime and dissuading others from joining. The current separation between aboriginal whaling (deemed permissible) and all other forms of harvesting (deemed impermissible) is hard to justify as a means of conserving whale stocks. It places unnecessary and counter-productive restrictions on sustainable development initiatives desired by a variety of communities interested in subsistence and artisanal (localized, family-based) whaling.[15]

As a result of extensive discussions, the social scientists concluded that small-scale whaling should be permissible on grounds of sustainability and equity "when it secures historically based practices of socially defined human groups that value whaling activities on a multi-dimensional basis."[16] They further found that sustainable and equitable whaling is most likely advanced when users and managers recognize the importance of "territoriality and the maintenance of social institutions that effectively restrict user access to commonly valued, used, and managed resources."[17] To identify when and where these conditions exist, the group suggested that five criteria or questions be used:

1. Is whaling conducted within *socially defined groups*?
2. Is whaling conducted within *identified territorial limits*?
3. Are whaling practices *socially reproducible over time*?
4. Are whaling practices *valued multi-dimensionally*?
5. Can management regimes ensure *biological sustainability*?

These questions are useful both in the Greenlandic context and elsewhere precisely because they move beyond problematic categories currently in place and focus instead on the issue of sustainability. In Greenland, whaling is conducted by *socially defined groups*, with members who "share a common culture or social bonds whose maintenance is demonstrably dependent upon whale harvesting or the consumption of whale products."[18]

The second question posed focuses on territoriality and the *existence of identifiable territorial limits*. Whaling, in particular, is "associated with specific shore-based communities."[19] As we have seen, Greenlandic whaling is community based. Whaling permits are assigned locally by municipal officials, who presumably are attuned to the needs of local communities and the ability of hunters to procure whales. Whale products are used strictly for local consumption, and no whale products are exported.

Greenlandic whaling also involves practices that are *socially reproducible over time*. These practices are based on rules and knowledge about whaling that is normally "handed down from generation to generation within the same community."[20] Younger hunters typically learn about whaling by working alongside experienced elders. These elders are often the modern-day version of a *piniartorsuaq* (great hunter), long recognized in Greenlandic communities as being productive, skilled, and knowledgeable providers and repositories of a wealth of knowledge.

A fourth question focuses on whether societies *value whaling multi-dimensionally*; that is, to what extent does a community have historical, economic, cultural, and nutritional relationships to whales and other marine resources? As this study reveals, ocean resources have long been the basis for the mixed economy in Greenlandic communities. Whaling is embedded within sociocultural and economic systems extending back many generations. It provides for nutritional needs, sustains indigenous social and cultural systems, and affirms Greenlandic identity. It also generates the cash necessary for sustaining livelihoods in communities that have few other economic alternatives.

A fifth and crucial question focuses on *biological sustainability*. The history of common property resource management shows clearly that biological research and monitoring systems must be in place to take into account the status of specific whale stocks and the environmental changes affecting them. Young and others suggest that

> the frequency and extent of such monitoring should be decided on a case-by-case basis. Given the inexact nature of fishery science, the best scientific judgments must necessarily allow for some (minimal) level of continuing disagreement among competent scientists regarding stock assessments.[21]

As this study shows, Greenlanders have made significant advances in monitoring and regulating whale catches, particularly since the advent of Home Rule. Biologists in Greenland and Denmark, working closely with those in other IWC member nations, have expanded considerably our knowledge about whale stocks in the North Atlantic. Greenlanders have also cooperated to improve the efficiency of the kill and to educate hunters about new regulations and quotas. Although the system is not perfect, Greenland's co-management regime has come a long way toward accommodating changing international expectations while also enabling local hunters to meet community needs and to create new whaling traditions.

Thinking globally, acting locally: Whaling in Greenland's future

When it comes to whaling and the use of other marine mammals, Greenlanders abide by the principle of thinking globally and acting locally. Although Greenland's ties to the world economy are increasing, many Greenlanders continue to identify themselves first and foremost as hunters and fishers. At a time when many people in industrialized countries are beginning to appreciate the importance of being rooted in a place and of being stewards of local resources, Greenlanders can draw upon four thousand years of history as a coastal people reliant on renewable resources from the sea.

As Greenland's population grows and as Greenlanders continue to become accustomed to a higher standard of living, they will undoubtedly seek new opportunities to create jobs and expand ties with other countries. Even with these changes, however, many Greenlanders—particularly those living in smaller, outlying settlements—will continue to rely on whales and other marine resources for nutritional, socio-cultural, and economic needs. Greenlanders realize that survival in the Arctic today requires thinking globally—about marine pollution, global warming, and nuclear proliferation—but survival also requires acting locally, being careful stewards and respectful users of renewable resources. Greenlanders have demonstrated their willingness to cooperate with the IWC in matters of resource conservation. They have accepted a zero quota on the take of both bowhead and humpback whales because of concerns about stock levels. They have done so even though there is a long history of taking both species in Greenland. They have also worked diligently to improve whaling practices in response to concerns about humane killing, and Home Rule officials continue to work with KNAPK and hunters to improve hunting practices and to gain increased compliance with whaling regulations.

In Greenland today, there need not be a contradiction between protecting the health of whale stocks and meeting the essential needs of human beings. Whaling can be a major contributor to sustainable development in Greenlandic communities; however, to achieve this goal there must be greater understanding of the roots of conflicting idea-systems about whaling, there must be greater awareness of the dynamics of mixed subsistence-cash economies, and there must be an effective management regime in place to monitor whale stocks and to ensure hunter compliance. Perhaps most importantly, there must also be recognition of the fact that human societies relate to their environments in different ways. Just as maintenance of biodiversity is vital to our collective future, so, too, is cultural diversity a source of continuing strength. Out of the richness and diversity of human societies may well come solutions to some of our most pressing problems—both in the Arctic and beyond.

Notes

Introduction (PP. 1–14)

1. In English, "Greenlanders' Land"; in Danish, Grønland.

2. World Commission on Environment and Development, *Our Common Future* (Oxford and New York: Oxford University Press, 1987).

3. See for example M. M. R. Freeman, "The International Whaling Commission, small-type whaling, and coming to terms with subsistence," *Human Organization* 53 (1993): 243–51; Douglas S. Butterworth, "Commentary: Science and sentimentality," *Nature* 357 (June 18, 1992): 532–34; Ray Gambell, "International management of whales and whaling: An historical review of the regulation of commercial and aboriginal subsistence whaling," *Arctic* 46, no. 2 (1993): 97–107; Patricia Birnie, "International legal issues in the management and protection of the whale: A review of four decades of experience," *Natural Resources Journal* 29 (fall 1989): 903–34.

4. O. R. Young and G. Osherenko, eds., *Polar politics: Creating international environmental regimes* (Ithaca: Cornell University Press, 1993), 1.

5. For an introduction to this rapidly growing literature, see Fikret Berkes, ed., *Common property resources: Ecology and community-based sustainable development* (London: Belhaven Press, 1989); F. Berkes, Peter George, and Richard Preston, "Co-management: The evolution of the theory and practice of joint administration of living resources" (McMaster University, Hamilton, TASO Research Report, Second Series, no. 1, 1991); H. P. Huntington, *Wildlife management and subsistence hunting in Alaska* (Seattle: University of Washington Press, 1992); Evelyn Pinkerton, ed., *Cooperative management of local fisheries: New directions for improved management and community development* (Vancouver: University of British Columbia Press, 1989); Gail Osherenko, *Sharing power with native users: Co-management regimes for native wildlife* (Ottawa: Canadian Arctic Resources Committee, Policy Paper 5, 1988).

6. Nancy C. Doubleday, "Arctic whales: Sustaining indigenous peoples and conserving Arctic resources," in *Elephants and whales: Resources for whom?* ed. M. M. R. Freeman and U. Kreuter (Basel: Gordon and Breach Science Publishers, 1994), 241–61; G. Osherenko and O. Young, *The age of the Arctic: Hot conflicts and cold realities* (Cambridge: Cambridge University Press, 1989); Osherenko, *Sharing power with native users*.

7. See Finn Gad, *The history of Greenland*, vol. 1, *Earliest times to 1700* (London: C. Hurst & Co., 1970), *The history of Greenland*, vol. 2, *1700–1782* (London: C. Hurst & Co., 1973), and "Grønlands historie: En oversigt fra ca. 1500 til 1945," *Grønl. Selsk. Skr.* 4 (1946).

8. Jens Dahl, *Arktisk selvstyre* (Copenhagen: Akademisk Forlag, 1986).

9. For a useful but controversial overview of Greenland's economy, see Martin Paldam, *Grønlands økonomiske udvikling: Hvad skal der til for at lukke gabet?* (Aarhus: Aarhus University Press, 1994).

10. *Sermitsiak'*, June 24, 1990, 14.

11. Mark Nuttall, *Arctic homeland: Kinship, community and development in northwest Greenland* (London: Belhaven Press, 1992).

12. Gambell, "International management."

13. G. Hardin, "The tragedy of the commons," *Science* 162 (1968): 1243–48.

14. J. N. Tønnesen and A. O. Johnsen, *The history of modern whaling* (Berkeley: University of California Press, 1982).

15. See J. Acheson, "Management of common-property resources," in *Economic anthropology*, ed. S. Plattner (Stanford: Stanford University Press, 1989), 357.

16. D. W. Bromley, ed., *Making the commons work: Theory, practice and policy* (San Francisco: ICS Press, 1992); E. Goldsmith, N. Hildyard, P. Bunyard, P. McCully, eds., "Whose common future?: A special issue," *The Ecologist* 22, no. 4 (1992): 122–210; D. Feeny, Fikret Berkes, Bonnie McCay, and James Acheson, "The tragedy of the commons: Twenty-two years later," *Human Ecology* 18, no. 1 (1990): 1–19; E. Ostrom, *Governing the commons: The evolution of institutions for collective action* (New York: Cambridge University Press, 1990); Berkes, *Common property resources*; Pinkerton, *Cooperative management*; P. Jacobs and D. A. Munro, *Conservation with equity: Strategies for sustainable development* (Gland: IUCN, 1987); Bonnie McCay and J. Acheson, eds., *The question of the commons: The culture and ecology of communal resources* (Tucson: The University of Arizona Press, 1987).

17. Acheson, "Common-property resources," 358.

18. Peter J. Usher, *The devolution of wildlife management and the prospects for wildlife conservation in the Northwest Territories* (Ottawa: Canadian Arctic Resources Committee, 1986).

19. D. M. Warren, L. J. Slikkerveer, D. Brokenshaw, eds., *Indigenous knowledge systems: The cultural dimension of development* (London: Zed Books, 1980); J. T. Inglis, ed., *Traditional ecological knowledge: Concepts and cases* (Ottawa: Canadian Museum of Nature and International Development Research Centre, 1993).

20. Berkes, George, and Preston, "Co-management."

21. Berkes, *Common property resources*, 26.

22. Alf Håkon Hoel, "Regionalization of international whale management: The case of the North Atlantic Marine Mammal Commission," *Arctic* 46, no. 2 (1993): 116–23.

23. Pinkerton, *Co-operative management*.

24. J. Dryzek and O. R. Young, "Internal colonialism in the circumpolar North," *Development and Change* 16 (1985): 123–45.

25. See Michael Pretes, "Underdeveloping the Arctic: Dependency, development and environmental control," in *Development perspectives for the 1990s*, ed. R. Prendergast and H. W. Singer (London: MacMillan, 1988); Michael Pretes, "Underdevelopment in two norths: The Brazilian Amazon and the Canadian Arctic," *Arctic* 41, no. 2 (1988): 109–16.

26. Harvey Feit, "Conflict areas in the management of renewable resources in the Canadian north: Perspectives based on conflicts and responses in northern Quebec" (paper presented at the Third National Workshop on People, Resources, and the Environment North of 60°, Yellowknife, June 1–3, 1983, Ottawa: Canadian Arctic Resources Committee, 1983), 404.

27. James Scott, *Weapons of the weak: Everyday forms of peasant resistance* (New Haven and London: Yale University Press, 1985).

28. Feit, "Conflict areas," 404.

29. Ibid., 405.

30. Martin Staniland, *What is political economy? A study of social theory and underdevelopment* (New Haven and London: Yale University Press, 1985).

31. Feit, "Conflict areas," 404.

32. Michael Redclift, *Development and the environmental crisis: Red or green alternatives?* (London and New York: Methuen, 1984), 130.

33. Feit, "Conflict areas," 373.

34. M. Cronon, *Changes in the land: Indians, colonists and the ecology of New England* (New York: Hill and Wang, 1983), ix.

1. History and political economy in Greenland (pp. 17–48)

1. Eric Wolf, *Europe and the people without history* (Berkeley: University of California Press, 1982), 3.

2. Robert McGhee, "The relationship between the mediaeval Norse and Eskimos," in *Between Greenland and America: Cross-cultural contacts and the environment in the Baffin Bay Area*, ed. L. Hacquebord and R. Vaughan (Groningen: University of Groningen, Arctic Centre, 1987).

3. Gwyn Jones, *The Norse Atlantic sagas*, 2d ed. (Oxford: Oxford University Press, 1986), 1.

4. Hugh Brody, "Ecology, politics and change: The case of the Eskimo," *Development and Change* 9 (1978): 31.

5. V. Stefansson (1913), quoted in Hugh Brody, *Living Arctic: Hunters of the Canadian North* (London and Boston: Faber and Faber), 9.

6. Gísli Palsson, *Coastal economies, cultural accounts: Human ecology and Icelandic discourse* (Manchester and New York: Manchester University Press, 1991).

7. B. Bender and B. Morris, "Twenty years of history, evolution and social change in gatherer-hunter studies," in *Hunters and gathererers*, vol. 1, *History, evolution and social change*, ed. T. Ingold, D. Riches, and J. Woodburn (Oxford: Berg Publishers, Limited, 1991), 10.

8. N. H. Bird-David, "Hunters-gatherers and other people: A re-examination," in ibid., 19.

9. Wolf, *Europe*; see also T. N. Headland and L. A. Reid, "Hunters-

gatherers and their neighbors from prehistory to the present," *Current Anthropology* 30, no. 1 (1989): 43–66.

10. Wolf, *Europe*, 18.

11. Adam Kuper, *The invention of primitive society: Transformations of an illusion* (London: Routledge, 1988).

12. David Damas, "Social anthropology of the central Eskimo," *Canadian Review of Sociology and Anthropology* 12, no. 3 (1975): 252–66.

13. Cited in Kate Sanderson, "Grind–Ambiguity and pressure to conform: Faroese whaling and the anti-whaling protest," in Freeman and Kreuter, *Elephants and whales*, 187.

14. Wolf, *Europe*, 75.

15. Ibid., 390.

16. Immanuel Wallerstein, *Unthinking social science: The limits of nineteenth-century paradigms* (Cambridge: Polity Press, 1991), 266–67.

17. See Michael Redclift, *Sustainable development: Exploring the contradictions* (London and New York: Methuen, 1987), and *Development and the environmental crisis.*

18. Wallerstein, *The limits of nineteenth-century paradigms*, 266–67.

19. Ibid.

20. M. Thompson and M. Warburton, "Uncertainty on a Himalayan scale," *Mountain Research and Development* 5 (1985): 115; see also Francis Sandbach, *Environment, ideology and policy* (Oxford: Basil Blackwell, 1980).

21. Wallerstein, *The limits of nineteenth-century paradigms*, 268; see also Redclift, *Sustainable development*; P. Marchak, "Canadian political economy," *Canadian Review of Sociology and Anthropology* 22, no. 5 (1985): 673–709; Carolyn Merchant, *The death of nature* (San Francisco: Harper and Row, 1980).

22. Redclift, *Sustainable development*, 124.

23. R. F. Murphy and J. H. Steward, "Tappers and trappers: Parallel processes in acculturation," *Economic Development and Cultural Change* 4 (1956): 336.

24. Karl Marx, *Capital*, vol. 1 (New York: International Publishers, 1967).

25. J. Nash, "Ethnographic aspects of the world capitalist system," *Annual Review of Anthropology* 10 (1981): 393–423; Tim Ingold, "Notes on the foraging mode of production," in Ingold, Riches, and Woodburn, *Hunters and gatherers*; A. Foster-Carter, "The modes of production controversy," *New Left Review* 107 (1978): 47–77; Barry Hindness and Paul Hirst, eds., *Pre-capitalist modes of production* (London: Routledge and Kegan Paul, 1975).

26. Wolf, *Europe*, 85–86.

27. Marshall Sahlins, *Stone age economics* (New York: Aldine Publishing Company, 1972).

28. Wolf, *Europe*, 76.

29. Ibid., 297.

30. Christian Berthelsen, I. Mortensen, and E. Mortensen, eds., *Kalaallit Nunaat/Greenland Atlas* (Nuuk: Pilersuiffik, 1990).

31. Thomas McGovern, "Climate, correlation and causation in Norse Greenland," *Arctic Anthropology* 28, no. 2 (1991): 77–100.

32. B. Grønnow and M. Meldgaard, "Boplads i dybfrost: fra Christianshåb Museums udgravninger på vestgrønlands ældste boplads," *Naturens Verden* (1988): 409–440.

33. Christian Vibe, "Arctic animals in relation to climatic fluctuations," *Meddelelser om Grønland* 170 (1967): 221–54.

34. Dahl, *Arktisk selvstyre.*

35. William W. Fitzhugh, "Paleo-Eskimo cultures of Greenland," in *Handbook of North American Indians: Arctic*, ed. D. Damas (Washington, D.C.: The Smithsonian Institution, 1984).

36. McGhee, "Mediaeval Norse and Eskimos."

37. Ibid., 57.

38. R. Petersen, E. Lemche, and F. O. Kapel, "Subsistence whaling in Greenland" (paper submitted to the International Whaling Commission Technical Committee Working Group on Subsistence Whaling, Histon, Cambridgeshire, UK), 5–6.

39. H. C. Gulløv, "Whales, whalers and Eskimos: The impact of European whaling on Eskimo society in West Greenland," in *Cultures in contact: The impact of European contacts on Native American Institutions, A.D. 1000–1800*, ed. W. W. Fitzhugh (Washington, D.C., and London: The Smithsonian Institution Press, 1985).

40. Finn Lynge, personal communication.

41. Finn Gad, "Grønland," in *Politikens Danmarkshistorie* (Copenhagen: Politiken, 1984).

42. G. C. Amdrup, L. Bobe, A. S. Jensen, and H. P. Steensby, eds., "Grønland i tohundredaaret for Hans Egedes Landing," *Meddelelser om Grønland* 60–61 (1921).

43. Gad, "Grønland," 557–58.

44. Egede named this "Håbets Ø," or Island of Hope.

45. H. C. Petersen, *Grønlændernes historie fra urtiden til 1925* (Nuuk: Atuakkiorfik, 1991).

46. Ibid.

47. H. Kleivan, "Contemporary Greenlanders," in Damas, *Handbook*, 525.

48. Petersen, *Grønlændernes historie*, 35.

49. Gad, "Grønland," 569.

50. See Knud Oldendow, "Grønlændervennen Hinrich Rink: Videnskabsmand, skribent, og grønlandsadministrator," *Det grønlandske Selskabs Skrifter* 18 (1955).

51. H. Rink, quoted in Petersen, *Grønlændernes historie*, 128.

52. Ibid.

53. E. L. B. Smidt, *Min tid i Grønland; Grønland i min tid* (København: Nyt Nordisk Forlag Arnold Busck, 1989).

54. J. Møller, quoted in Petersen, *Grønlændernes historie*, 249.

55. *Sonja* was built in England but used primarily in Antarctic whaling by the Norwegians.

56. Petersen, *Grønlændernes historie*, 170.

57. Kleivan, "Contemporary Greenlanders."

58. Koch, quoted in ibid., 702.

59. Dahl, *Arktisk selvstyre.*

60. Kleivan, "Contemporary Greenlanders."

61. Ibid.

62. Dahl, *Arktisk selvstyre*, 21.

63. Ibid., 50.
64. Kleivan, "Contemporary Greenlanders," 715.
65. Dahl, *Arktisk selvstyre*.
66. Grønlands Hjemmestyre, *Grønland/Kalaallit Nunaat 1995. Statistisk årbog/Ukiumoortumik kisitsisitigut passissutissat* (Nuuk: Grønlands Statistik/Atuakkiorfik, 1995).
67. G. Poole, M. Pretes, and K. Sinding, "Managing Greenland's mineral revenues," in *Nordic Arctic research on contemporary Arctic problems: Proceedings from Nordic Arctic Research Forum Symposium 1992*, ed. L. Lyck (Aalborg: Aalborg University Press, 1992).
68. Advisory Committee for Greenland's Economy [Det Rådgivende Udvalg Vedrørende Grønlands Økonomi], "Beretning om den økonomiske udvikling i Grønland 1993" (Copenhagen, July 1994).
69. Paldam, *Grønlands økonomiske udvikling*, 75.
70. Ibid.
71. Ibid.
72. Ibid., 58.
73. Ibid.
74. Ibid., 80.
75. Grønlands Hjemmestyre, *Grønland/Kalaallit Nunaat 1995*.
76. Poole, Pretes, and Sinding, "Greenland's mineral revenues"; and K. Sinding, "At the crossroads: Mining policy in Greenland," *Arctic* 45, no. 3 (1992): 226–32.
77. *Atuagagdliutit* 10, April 28, 1994, 10.
78. Ibid.
79. P. A. Andersen, *Sermitsiak'* 20, May 18, 1990, 14.
80. Letter to the editor, *Sermitsiak'* 20, May 18, 1990, 13.
81. *Sermitsiak'*, June 22, 1990, 16.

2. Changing production relations in Greenland (pp. 49–74)

1. McCay defines simple commodity production as that "based on relatively small-scale, simple technology; work groups organized around kinship, friendship, or temporary collegiality but with little difference between owners and laborers; widespread sharing of costs, risks, benefits, and windfalls; and a variable subsistence/market allocation of production." See Bonnie McCay, "Development issues in fisheries as agrarian systems," *Culture and Agriculture* 11 (1981): 1–8.
2. See Robert Wolfe and Linda J. Ellanna, "Resource use and socioeconomic systems: Case studies of fishing and hunting in Alaskan communities" (Alaska Department of Fish and Game, Division of Subsistence, Juneau, 1983); and Oran R. Young, "The mixed economies of village Alaska," in *Arctic politics: Conflict and cooperation in the circumpolar North* (Hanover: University Press of New England, 1992).
3. In Danish, Diskofjord.
4. Qeqertarsuaq means "big island" in Greenlandic; in Danish, Disko Ø.

5. O. Rosing, "Diskomik oqaluttualiaq" in *K'ek'ertarssuak'/Godhavn*, ed. J. Fisker (Umanak: Nordiske Landes Bogforlag, 1984).

6. Amdrup et al., "Hans Egedes Landing."

7. Fisker, *K'ek'ertarssuak'/Godhavn*.

8. P. P. Sveistrup and S. Dalgaard, "Det danske styre af Grønland, 1825–1850," *Meddelelser om Grønland* 145, no. 1 (1945).

9. Poul Møller and Susanne Dybbroe, "Fanger/fisker-Lønarbejder? En undersøgelse af sammenhængene mellem fangst og lønarbejde i Godhavn, 1976 & 1977" (Institut for forhistorisk arkæologi, middelalder-arkæologi, etnografi og socialantropologi, Århus Universitet, Århus, 1981).

10. Finn Kapel, "Catch of minke whales by fishing vessels in West Greenland," *Rep. int. Whal. Commn.* 28 (1978): 217–26.

11. Data presented here were gathered by the author while in Qeqertarsuaq in 1989–1990, 1992, 1994, and 1995.

12. See Richard A. Caulfield, "*Qeqertarsuarmi arfanniarneq*: Greenlandic Inuit whaling in Qeqertarsuaq Kommune, West Greenland" (report submitted to 43d Annual Meeting of the International Whaling Commission, Reykjavik, Iceland, May 1991, TC/43/AS4).

13. Field data 1995.

14. Caulfield, "Greenlandic Inuit whaling in Qeqertarsuaq Kommune."

15. Ibid.

16. Ibid.

17. Finn O. Kapel and Robert Petersen, "Subsistence Hunting—the Greenland Case," *Rpt. int. Whal. Commn.*, Special Issue 4 (1984): 51–71.

18. Caulfield, "Greenlandic Inuit whaling in Qeqertarsuaq Kommune."

19. Ibid.

20. Greenland Home Rule Government, "Fisker- og fangerfamiliers levevilkår. Rapport nr. 3 fra undersøgelsen af befolkningens økonomiske og materielle levevilkår" (Grønlands Statistiske Kontor, Nuuk, 1995).

21. Lee Guemple, "Teaching social relations to Inuit children," in *Hunters and gatherers*, vol. 2, *Property, power and ideology*, ed. Tim Ingold, D. Riches, and J. Woodburn (Oxford: Berg Publishers Limited, 1988), 148.

22. Wolf, *Europe*.

23. R. J. Wolfe, J. J. Gross, S. J. Langdon, J. M. Wright, G. K. Sherrod, L. J. Ellanna, V. Smith, and P. J. Usher, "Subsistence-based economies in coastal communities of southwest Alaska" (Alaska Department of Fish and Game, Division of Subsistence and US Department of the Interior, Minerals Management Service, Technical Paper no. 89, Juneau, 1984), 380.

24. For related examples elsewhere in the Arctic, see George Wenzel, *Animal rights, human rights: Ecology, economy and ideology in the Canadian Arctic* (London: Belhaven Press, 1991); and Barbara A. Bodenhorn, "The animals they come to me, they know I share: Iñupiat kinship, changing economic relations and enduring world view on Alaska's North Slope" (Ph.D. diss., Department of Social Anthropology, University of Cambridge, 1991).

25. S. E. Larsen and K. G. Hansen, "Inuit and whales at Sarfaq (Greenland)" (paper presented to the 42d Annual Meeting of the International Whaling Commission, Noordwijk, The Netherlands, July 2–6, 1990).

26. Field notes, Qeqertarsuaq, October 28, 1989.

27. Robert Petersen, "On traditional and present distribution channels in the subsistence hunting in Greenland" (paper presented at the 1988 Inuit Studies Conference, Copenhagen, Denmark).

28. Inge Kleivan, "West Greenland before 1950," in Damas, *Handbook.*

29. Per Langgård, "Modernization and traditional interpersonal relations in a small Greenlandic community: A case study from southern Greenland," *Arctic Anthropology* 23, nos. 1–2 (1986): 299–314.

30. Dahl, *Arktisk Selvstyre,* and "Tradition og kultur i den grønlandske naturudnyttelse," *Grønland* 10 (1987): 295–304.

31. Caulfield, "Greenlandic Inuit whaling in Qeqertarsuaq Kommune."

32. Wolfe et al., "Subsistence-based economies."

3. Vessels, kin, and harpoons (pp. 77–110)

1. Goldsmith et al., "Whose common future?" 203.

2. Judith E. Zeh, Christopher W. Clark, John C. George, David Withrow, Geoffrey M. Carroll, and William R. Koski, "Current population size and dynamics," in *The bowhead whale,* ed. J. J. Burns, J. J. Montague, C. J. Cowles (Lawrence, Kans.: The Society for Marine Mammology, 1993), 409–89.

3. J. Sigurjonsson, "Whale stocks off Iceland," *North Atlantic Studies* 2, nos. 1–2 (1990): 64–76.

4. Joseph Horwood, *Biology and exploitation of the minke whale* (Boca Raton: CRC Press, Inc., 1990); and Eric W. Born, "Aspects of present-day maritime subsistence hunting in the Thule Area, Northwest Greenland," in Hacquebord and Vaughn, *Between Greenland and America,* 109–32.

5. Finn O. Kapel, "Sex ratio and seasonal distribution of catches of minke whales in Greenland," *Rep. int. Whal. Commn.* 30 (1980): SC/31/Doc 9.

6. Greenland Natural Resources Institute [Pinngortitaleriffik], "Kalaallit Nunaanni miluumasut imarmiut: Sulissutiginniffiunerat aamma ilisimatusarneq, 1. januar 1995 qanoq issuusiat" (Greenland Natural Resources Institute, Nuuk, 1995).

7. Ibid.

8. See Butterworth, "Commentary."

9. Finn O. Kapel, "Exploitation of large whales in West Greenland in the twentieth century," *Rep. int. Whal. Commn.* 29 (1979): SC/30/Doc 23.

10. Greenland Natural Resources Institute [Pinngortitaleriffik], "Kalaallit Nunaanni miluumasut imarmiut."

11. IWC, "Report of the Scientific Committee," IWC/47/4 (Annex F), 1995.

12. S. Katona and J. Beard, "Population size, migrations and substock structure of the humpback whale (*Megaptera novaeangliae*) in the western North Atlantic Ocean," *Rep. int. Whal. Commn.,* special issue 12 (1990).

13. Greenland Natural Resources Institute [Pinngortitaleriffik], "Kalaallit Nunaanni miluumasut imarmiut."

14. H. C. Petersen, personal communication, 1992.

15. B. Grønnow, B. Meldgaard, and M. Meldgaard, "Boplads i dybfrost: fra Christianshåb Museums udgravninger på vestgrønlands ældste boplads," *Naturens Verden* (1988): 409–40.

16. Gad, *The history of Greenland*, vol. 1; H. C. Petersen, *Skinboats of Greenland* (Roskilde: The National Museum of Denmark, The Museum of Greenland, and The Viking Ship Museum in Roskilde).

17. Otto Fabricius, "Fauna groenlandica. Mammalia et aves," trans. O. Helms, *Grønl. Selsk. Skr.* 6 (1780), (1929), and "Om stubhvalen," *Da. Vidn. Selsk. Skr.* 6, no. 1 (1809); Heinrich Rink, *Danish Greenland, its people and products* (1877; reprint, London: C. Hurst & Co. and A. Busch, 1924).

18. H. C. Petersen, *Skinboats of Greenland*.

19. R. Petersen, "Fangst og bosættelse: Studier fra to grønlandske fangersamfund" (manuscript at Institute of Eskimology, University of Copenhagen).

20. See Nuttall, *Arctic homeland*; R. M. Søby, "The Eskimo animal cult," *Folk* 11–12 (1969/70): 43–78; I. Kleivan, "Mitaartut: Vestiges of the Eskimo sea-woman cult in West Greenland," *Meddelelser om Grønland* 161, no. 5 (1960).

21. Lars Dalager, "Grønlandske Relationer . . . 1752," ed. L. Bobe, *Grønl. Selsk. Skr.* 2 (1915): 56.

22. H. C. Glahn, "Dagbøger 1767–68," *Grønl. Selsk. Skr.* 4 (1921).

23. Ibid.

24. Dalager, "Grønlandske Relationer," 87.

25. Gulløv, "Whales, whalers and Eskimos."

26. H. V. Bang, "Om rationel hvalfangst i Grønland," *Grønl. Selsk. Aarsskr.* (1912): 19–41.

27. H. C. Petersen, *Skinboats of Greenland*.

28. R. Vaughan, "Bowhead whaling in Davis Strait and Baffin Bay during the 18th and 19th centuries," *Polar Record* 23, no. 144 (1986): 289–300.

29. Rink, *Danish Greenland*.

30. Fisker, *K'ek'ertarssuak'/Godhavn*; H. Winge, "Grønlands pattedyr," *Meddelelser om Grønland* 21, no. 2 (1902): 317–521.

31. Bang, "Om rationel hvalfangst i Grønland."

32. Winge, "Grønlands pattedyr."

33. H. C. Petersen, *Grønlændernes historie*.

34. Winge, "Grønlands pattedyr"; Adam Broberg, "K'ekertarssuarme piniartorssusimasok," *Avangnamiok'* 1 (1945): 3–5; G. Kleist, "Oqalualaarutit," *Atuagagdliutit* 8, January 14, 1898, 113–17; Hans Peter Grønvold, "Qeqertarsuarmiu arfanniaq" [the whaler from Qeqertarsuaq], (unpublished manuscript, Qeqertarsuaq, Greenland).

35. Broberg, "K'ekertarssuarme piniartorssusimasok."

36. Kleist, "Oqalualaarutit."

37. Broberg, "K'ekertarssuarme piniartorssusimasok," 3–5. Translation by M. Berthelsen and the author.

38. Bang, "Om rationel hvalfangst i Grønland."

39. Jens Dalager, "Atuagagliutinut ilángutagssiaq, 1896-me," *Atuagagdliutit* 1, October 1, 1897, 1–6.

40. Kapel, "Exploitation of large whales."

41. See ibid.; Winge, "Grønlands pattedyr"; Ministry for Greenland,

"Grønlændernes fangst af søpattedyr: Knølhvalen," *Ber. vedr. Grønl. Styrelse* 1 (1944); John Møller, "Knølhvalfangst," in *Grønlandske Fangere Fortæller*, ed. K. Hansen (København: Nordiske Landes Bogforlag, 1971); Sveistrup and Dalgaard, "Det danske styre af Grønland."

42. Knud Oldendow, "Naturfredning i Grønland," *Grønl. Selsk. Skr.* 9, 1935.

43. Sveistrup and Dalgaard, "Det danske styre af Grønland."

44. Kapel, "Exploitation of large whales."

45. H. C. Petersen, *Skinboats of Greenland.*

46. See Bang, "Om rationel hvalfangst i Grønland."

47. Kapel, "Exploitation of large whales."

48. R. Tving, "Skibe i Grønlands-farten," *Grønland* (1955): 38–40.

49. R. Tving, "M/S Sværdfisken," *Grønland* (1955): 238–39.

50. Kapel, "Exploitation of large whales."

51. Smidt, *Min tid i Grønland*, 125.

52. Kapel, "Exploitation of large whales."

53. Ibid.

54. Smidt, *Min tid i Grønland.*

55. Kapel, "Exploitation of large whales," 198.

56. Smidt, *Min tid i Grønland.*

57. Kapel, "Catch of minke whales."

58. Grønlands Landsråd, *Grønlandslandsrådforhandlinger* (Nuuk/Godthåb: Sydgrønlands Bogtrykeri, 1948), 128.

59. Alipak Johansen, "Auvek aitsat taima angitigissumik pissakarpok. [Auveq just had a large catch], *Avangnamiok'* 1 (1953): 1.

60. Kapel, "Catch of minke whales."

61. Ibid.

62. Grønlands Landsråd, *Grønlandslandsrådforhandlinger, 1971 (efterår)*, 396–98.

63. Ibid.

64. Finn Breinholt-Larsen, personal communication, 1990.

65. Compiled from Greenland Home Rule Government, "Redegørelse om fangererhvervet 1994," and *Grønland/Kalaallit Nunaat 1990. Statistisk årbog/ Ukiumoortumik kisitsisitigut passissutissat.* See also Horwood, *Biology of minke whale.*

66. Kapel, "Catch of minke whales," 220.

67. Caulfield, "Greenlandic Inuit whaling in Qeqertarsuaq Kommune."

68. Ibid.

69. Greenland Home Rule Government, "Greenland action plan on whale hunting methods, 1995," IWC report no. TC/47.

70. Ibid.

71. Ibid.

72. Caulfield, "Greenlandic Inuit whaling in Qeqertarsuaq Kommune."

73. Ibid.

74. Ibid.

75. R. Petersen, "Distribution channels in Greenland."

76. Greenland Home Rule Government, "Redegørelse om fangererhvervet 1994," 59.

77. Caulfield, "Greenlandic Inuit whaling in Qeqertarsuaq Kommune."

78. G. P. Donovan, ed., "Aboriginal/subsistence whaling," *Rep. int. Whal. Commn.*, Special Issue 4 (1982): 40.

79. Robert Petersen, "Cultural needs: Communal aspects of preparation for whaling, of the hunt itself and of the ensuing products," (report presented to the 39th Annual Meeting of the International Whaling Commission, Bournemouth, UK, June 22–26, 1987), TC/39/AS4.

80. Dahl, "Tradition og kultur."

4. Greenland's whaling regime (pp. 111–146)

1. Finn Lynge, *Arctic wars: Endangered animals, endangered peoples* (Hanover: University Press of New England, 1992).

2. See Thomas R. Berger, *A long and terrible shadow: White values, native rights in the Americas, 1492–1992* (Vancouver and Toronto: Douglas & McIntyre, 1991); and Nancy Doubleday, "Aboriginal subsistence whaling: The right of Inuit to hunt whales and implications for international environmental law," *Denver Journal of International Law and Policy* 17, no. 2 (1989): 373–94.

3. United Nations General Assembly, Resolution 2200/XXI, 1966.

4. International Work Group for Indigenous Affairs (IWGIA), *Arctic environment: Indigenous perspectives* (IWGIA Document 69, Copenhagen, August 1991).

5. For examples from an extensive and diverse literature, see Doubleday, "Arctic whales"; O. R. Young, M. M. R. Freeman, G. Osherenko, R. R. Andersen, R. A. Caulfield, R. L. Friedheim, S. J. Langdon, M. Ris, and P. J. Usher, "Commentary: Subsistence, sustainability, and sea mammals: Reconstructing the international whaling regime," *Ocean and Coastal Management* 23 (1994): 117–27; M. M. R. Freeman, "International Whaling Commission," and "Science and trans-science in the whaling debate," in Freeman and Kreuter, *Elephants and whales*; Ray Gambell, "International management," and "The International Whaling Commission—quo vadis?" *Mammal Review* 20 (1990): 31–43; Steinar Andreasen, "The effectiveness of the International Whaling Commission," *Arctic* 46 (1993): 108–15, and "Science and politics in the international management of whales," *Marine Policy* 13, no. 2 (1989): 99–117; A. H. Hoel, "The International Whaling Commission 1972–1984: New members, new concerns" (Fridtjof Nansens Institutet, Oslo, Study No. R:003-1985, 1985); M. J. Peterson, "Whalers, cetologists, environmentalists and the international management of whaling," *International Organization* 46, no. 1 (1992): 149–53; Patricia Birnie, "International legal issues," and *International regulation of whaling*, 2 vols. (New York: Oceana Publications, 1985); Sidney Holt, "Whale mining, whale saving," *Marine Policy* (July 1985), 192–213.

6. Gambell, "International management."

7. Ibid.

8. United Nations Convention on the Law of the Sea (UNCLOS), UN Document No. A/Conf. 62/122, 21 I.L.M. 1261, 1982.

9. Susan J. Buck, "Multi-jurisdictional resources: Testing a typology for problem-structuring," in Berkes, *Common property resources*, 128.

10. Cyril De Klemm, "Migratory species in international law," *Natural Resources Journal* 29 (1989): 941.

11. Gambell, "International management."

12. See Donovan, "Aboriginal/subsistence whaling."

13. Gambell, "International management."

14. Ibid.

15. Cited in Doubleday, "Aboriginal subsistence whaling," 386.

16. Gambell, "International management."

17. Ibid., 104.

18. Ibid.; and G. P. Kirkwood, "Background to the development of Revised Management Procedures," *Rep. int. Whal. Commn.* 42 (1992): 236–43.

19. G. P. Donovan, "The International Whaling Commission and the Revised Management Procedure," in *Additional Essays on Whales and Man*, ed. G. Blichtfeldt (Reine, Norway: High North Alliance, 1995).

20. These elements, together with a program for inspection and observation, are referred to as the "Revised Management Scheme."

21. See, for example, US Marine Mammal Commission, "Issues facing the International Whaling Commission and the United States regarding the resumption of commercial whaling and the future conservation of cetaceans," (Washington, D.C., December 5, 1991).

22. Gambell, "International management," 105.

23. IWC Resolution 1994-4, "Resolution on a Review of Aboriginal Subsistence Management Procedures," in IWC, 1995, "Chairman's Report of the Forty-Sixth Annual Meeting," Appendix 4, *Rep. int. Whal. Commn.* 45, 42–44.

24. "Extract from the Scientific Committee Report [IWC/47/4] for Consideration by the Working Group," May 29, 1995.

25. Gambell, "International management," 106.

26. Inuit Circumpolar Conference, "Circumpolar whaling and the ICC whaling agenda: A choice for Inuit to make" (report presented at the 1995 General Assembly of the Inuit Circumpolar Conference, Nome, Alaska, July 1995).

27. Ibid., 1.

28. Caleb Pungowiyi, "Arfanniarnermut kommissionimik naapitsineq ICC-p præsidentiata tupassutigisorujussuusimavaa," *Atuagagdliutit* 42, June 7, 1994, 10.

29. Doubleday, "Aboriginal subsistence whaling," 387.

30. Arne Kalland, "Aboriginal subsistence whaling: A concept in the service of imperialism?" in *Bigger than whales*, High North Alliance (special publication produced for the 44th Annual Meeting of the International Whaling Commission, Glasgow, Scotland, June 1992).

31. Jens Brøsted, "Urbefolkningsretten og den grønlandske hvalfangst," *Retfærd i Grønland* (1980): 62–73.

32. Ibid.

33. Ibid., 70.

34. Ibid., 71.

35. *Landstingslov om erhvervsmæssigt fiskeri, fangst og jagt*, October 13, 1980.

36. Nikolaj Heinrich, *Atuagagdliutit*, July 31, 1985, 3.

37. Greenland Home Rule Government, "Greenland Home Rule authorities and the monitoring of whale hunt in Greenland," (report presented to the 39th Annual Meeting of the International Whaling Commission, Bournemouth, UK, June 22–26, 1987).

38. Greenland Home Rule Government, *Hjemmestyrets bekendtgørelse nr. 18 af 22. juli 1993 om jagtbeviser* (Nuuk, July 22, 1993).

39. Greenland Home Rule Government, "Monitoring of whale hunt in Greenland."

40. Mads Fægtebord, "Inuit organizations and whaling policies," *North Atlantic Studies* 2, nos. 1–2, 127.

41. Greenland Home Rule Government, "Greenland subsistence hunting" (report prepared for the 41st Annual Meeting of the International Whaling Commission, San Diego, California, 1989), IWC/41/22.

42. Greenland Home Rule Government, "Implementation of the detonating grenade harpoon in Greenlandic whaling on an experimental basis" (report prepared for the 40th Annual Meeting of the International Whaling Commission, Auckland, New Zealand, May 30–June 3, 1988), TC/40/HK 4.

43. Hans Iversen, in Inuit Circumpolar Conference, *Inuit tusaataat/Inuit whaling*. Nuuk: ICC, 1992.

44. Caulfield field notes, Qeqertarsuaq Municipality, 1990.

45. Ibid.

46. Finn Larsen, Letter of May 12, 1991, to print media in Denmark (Copenhagen: Greenland Fisheries Research Institute); some twenty separate infractions were noted.

47. S. E. Larsen and K. G. Hansen, "Inuit and whales at Sarfaq," 14, TC/42/SEST 4.

48. Kaj Egede in *Sermitsiak'*, January 19, 1990, 26.

49. Amalie Jessen in *Sermitsiak'* 17, April 24, 1990, 3.

50. *Sermitsiak'* 20, May 20, 1994, 7.

51. *Sermitsiak'* 6, January 21, 1994, 32.

52. Ibid., 12.

53. Unpublished letter, 1994.

54. *Sermitsiak'*, October 25, 1991, 6.

55. Greenland Home Rule Government, "Aboriginal subsistence working group—Danish statement" (report presented to the 40th Annual Meeting of the International Whaling Commission), 40/AS/3. Auckland, NZ, 30 May–3 June 1988.

56. KNAPK letter to Greenland's Minister for Fisheries, Hunting, and Agriculture, January 26, 1994.

57. Amalie Jessen, Greenland Home Rule government, personal communication, 1996.

58. Kalland, "Aboriginal subsistence whaling."

59. Brian Moeran, "The cultural construction of value: 'Subsistence,' 'commercial' and other terms in the debate about whaling," in *The report of the symposium on utilization of marine living resources for subsistence, January 21–23, 1992, Taiji, Japan*, vol. 1, ed. B. Moeran (Tokyo: The Institute of Cetacean Research, 1992), 99. See also J. Parry and M. Bloch, eds., *Money and the morality of ex-*

change (Cambridge: Cambridge University Press, 1992); and Georg Simmel, *The philosophy of money*, trans. and ed. David Frisby and Tom Bottomore, 2d edition (London: Routledge, 1990).

60. Lynge, *Arctic wars*.

61. Nicholas Peterson, "Introduction: Cash, commoditization and changing foragers," in *Cash, commoditization and changing foragers*, ed. N. Petersen and T. Matsuyama (Osaka: National Museum of Ethnology, 1991), 2.

62. Ibid.

63. *Sermitsiak'* 12, March 16, 1990, 12.

64. E. Hobsbawm and T. Ranger, *The invention of tradition* (Cambridge: Cambridge University Press, 1983).

65. Quoted in Greenland Home Rule Government, "Greenland subsistence hunting," 11.

66. KNAPK letter to Greenland's Minister for Fisheries, Hunting, and Agriculture, January 26, 1994.

5. Initiatives on the periphery (pp. 147–166)

1. Scott, *Weapons of the weak*.

2. Goldsmith et al., "Whose common future?" 204.

3. Ibid.

4. For examples with differing perspectives, see Lynge, *Arctic wars*; Ö. Jónsson, ed., *Whales and ethics* (Reykjavik: Fisheries Institute and University Press, University of Iceland, 1992); A. D'Amato and S. Chopra, "Whales: Their emerging right to life," *The American Journal of International Law* 85, no. 1 (1991): 21–62; G. Wenzel, *Animal rights*; Robbins Barstow, "Beyond whale species survival: Peaceful coexistence and mutual enrichment as a basis for human/cetacean relations," *Sonar* 2 (Bath, UK, magazine of The Whale and Dolphin Society) (autumn 1989): 10–13; David Day, *The whale war* (London: Routledge & Kegan Paul, 1987); R. Keith and A. Saunders, eds., *A question of rights: Northern wildlife management and the anti-harvest movement* (Ottawa: Canadian Arctic Resources Committee, 1989).

5. Wenzel, *Animal rights*, 142.

6. Ibid.

7. Quoted in Ibid., 166.

8. Quoted in Ibid., 167.

9. D'Amato and Chopra, "Whales: Their emerging right to life," 59.

10. Mads Heide-Jørgensen, "Small cetaceans in Greenland: Hunting and biology," *North Atlantic Studies* 2, nos. 1–2 (1990): 55–58.

11. Amalie Jessen, "Modern Inuit whaling in Greenland" (paper presented to the First International Congress on Arctic Social Sciences, Quebec, Canada, October 28–31, 1992).

12. Heide-Jørgensen, "Small cetaceans in Greenland."

13. Greenland Home Rule Government, "Redegørelse om fangererhvervet 1994."

14. Quoted from Danmarks Radio television program, "Man skyder da hvaler," August 1991.

15. US Marine Mammal Commission, "Issues facing the International Whaling Commission."

16. Barstow, "Beyond whale species survival."

17. Freeman and Kreuter, *Elephants and whales*.

18. Lynge, *Arctic wars*.

19. Quoted from Danmarks Radio television program, "Man skyder da hvaler," August 1991.

20. Knud Rasmussen, *Intellectual culture of the Iglulik Eskimos: Report of the Fifth Thule Expedition, 1921–24*, vol. 7, no. 1 (1929), 56.

21. Inuit Circumpolar Conference, *Inuit tusaataat/Inuit whaling*, 4.

22. Ted L. McDorman, "GATT and international reaction to U.S. practices," in High North Alliance, *Additional essays on whales and man*. Reine, Norway: High North Alliance.

23. Ibid.

24. Inuit Circumpolar Conference, *Inuit tusaataat/Inuit whaling*, 4.

25. *Sermitsiak'* 20, May 20, 1994, 7.

26. Ibid.

27. *Sermitsiak'*, May 17, 1991, 1.

28. Quoted in *Inuit whaling, Inuit tusaataat* (June 1991), 4. Nuuk: ICC.

29. Ibid.

30. *Sermitsiak'*, May 31, 1991, 16.

31. Ibid., 2.

32. Danish Parliament (*Folketinget*), "Beretning angående hvalfangst" (København: Beretning nr. 3, Folketinget, 1990–1991 [2. samling], Beretning afgivet af Miljø- og Planlægnings-udvalget, May 24, 1991).

33. *Sermitsiak'* 7, June 1991, 12.

34. Lynge, *Arctic wars*.

35. *Politiken*, July 1, 1991, 2.

36. Ibid.

37. *Sermitsiak'* 20, May 20, 1994, 6.

38. "Greenland's whalers want to export whale meat," *High North News*, May 15, 1995, 8.

39. *Sermitsiak'* 20, May 20, 1994, 6.

40. Letter to Greenland's Minister for Fisheries, Hunting, and Agriculture, January 26, 1994.

41. NAMMCO (North Atlantic Marine Mammal Commission), "Agreement on cooperation in research, conservation and management of marine mammals in the North Atlantic" (signed in Nuuk, Greenland, April 9, 1992).

42. Hoel, "Regionalization."

43. Ibid.

44. Quoted in "Whaling: Samples from a Contemporary Debate," in E. Vestergaard, ed., *Whaling communities/North Atlantic Studies* 2, nos. 1–2 (1990).

45. Ibid., 121.

46. Hoel, "Regionalization."

47. United Nations Convention on the Law of the Sea (UNCLOS), UN Document No. A/Conf. 62/122, 21 I.L.M. 1261, 1982.
48. Quoted in *International Fishing News*, November 1992.
49. *The International Harpoon* 2 (1992): 4.
50. United Nations Conference on Environment and Development, *Agenda 21*, Article 17.1.
51. Ibid., Articles 17.46, 17.62, and 17.90.
52. Hoel, "Regionalization," 123.
53. Ibid.
54. Feit, "Conflict areas."
55. Mads Fægteborg, "Inuit Circumpolar Conference: An indigenous organization as an instrument of influence," in *Nordic Arctic research on contemporary Arctic problems*, ed. L. Lyck (Aalborg: Aalborg University Press, 1992).
56. Mads Fægteborg, "Inuit organizations," 126.
57. Quoted in Inuit Circumpolar Conference, "Circumpolar whaling and ICC whaling agenda," 1.
58. Inuit Circumpolar Conference, *Inuit tusaataat/Inuit whaling*, 10.
59. Inuit Circumpolar Conference, "Circumpolar whaling and ICC whaling agenda."
60. Ibid., 1.
61. Ibid., 21.
62. Ibid., 20.
63. Inuit Circumpolar Conference, *Inuit tusaataat/Inuit whaling*, 10.

Conclusion (pp. 167–178)

1. World Commission on Environment and Development, *Our common future*.
2. For a critique of this concept, see Frank Duerden, "A critical look at sustainable development in the Canadian North," *Arctic* 45, no. 3 (1991): 219–25; Redclift, *Sustainable development*.
3. Wolf, *Europe*, 76.
4. Berkes, *Common property resources*, 26.
5. Quoted in S. Ritter, "Cooperative wildlife management systems: A case study of whale management in Greenland" (paper prepared for Institute of Arctic Studies, Dartmouth College, Hanover, N.H., 1990), 32.
6. William Aron, "The commons revisited: Thoughts on marine mammal management," *Coastal Management* 16 (1988): 99–110.
7. Goldsmith et al., "Whose common future?"
8. Ibid., 181.
9. D'Amato and Chopra, "Whales: Their emerging right to life," 61.
10. M. Thompson and M. Warburton, "Uncertainty," 121.
11. Ibid., 122.
12. Ibid.
13. Goldsmith et al., "Whose common future?"
14. Young et al., "Commentary."

15. Ibid., 119.
16. Ibid.
17. Ibid.
18. Ibid., 120.
19. Ibid.
20. Ibid., 121.
21. Ibid.

Appendix

Atuagaq Kalaallit Nunaanni arfanniarneq pillugu eqikkarnera "Greenlanders, whales, and whaling" (Kalaallit, arferit arfanniarnerlu)

Richard A. Caulfield
University of Alaska, Fairbanks USA

Rikka Caulfield nuliaa meeraallu 1989-mit 1990-mut Qeqertarsuarmi najugaqarput. Qeqertarsuup kommunianit, Kalaallit Nunaanni Namminersornerullutik Oqartussanit aamma Qeqertarsuup Kommuniani KNAPP-mit akuersissuteqarfigineqarluni arfanniarneq pillugu misissuivoq. Atuagaq una misissuineq matumani allaatigaa. Caulfield 1995-imi 1996-imillu Nuummi Ilisimatusarfimmi sulisarpoq ilisimatuutut tikeraartutut. Caulfield University of Alaska Fairbanks-imi ilisimatuujuvoq.

Kalaallit Nunaanni inuit tamarluinnarmik imartamik pisuussutaat paasisimavaat, pingaartumik puisit arferillu, inuuniarnermut pingaaruteqarluinnartut. Kikkulluunniit, inuit peqatigiiffiillu Kalaallit Nunaata avataaneersut, arfanniarnerup pingaaruteqarneranik paasisimannittuunngitsut, sunniuteqariartuinnarput kalaallit imaani miluumasunik piniarnerannut. Pingaartumik Nunarsuarmi Arfanniarneq pillugu ataatsimiititaliaq (IWC) aqqutigalugu ukiuni kingusinnerusuni arfattassat amerlasusii ikilisinneqarput. Namminersornerullutik Oqartussat, KNAPK allallu arfanniarnermut akuliusimasut pisassat amerlassusissaannik isumaqatigiissuteqartarnerannut peqatigillugu pisariaqarpoq misissuisoqarnissaa arfanniarnerup pinngitsoorneqarsinnaanngineranik allaatiginnittartussanik.

Atuagaq manna tunngaveqarpoq piniartut najugarisami, pisortat allallu oqaloqatigalugit paasissutissiissutaasigut, ilaquttakka ilagalugit Qeqertarsuarmi najugaqarsimanitta nalaani. Piffissami tamatumani kalaallit kitaamiut oqaasii ilinniarsimavakka. Lars Pele Berthelsen-ilu suleqatigeqqissartarsimallugu qujamasuutigeqisannik misissuininni tapersersortigilluarsimagakka.

Misissuininni inuit Qeqertarsuarmi Kangerlummilu inoqutigiinnit 60-ineersut oqaloqatigaakka inuuniarnerat pillugu, aammalu oqaloqatigiissutigisarpavut arfanniarnerni peqataasarsimanerat. Nalunaarutip takutipaa arfernit arfanniarnernilu pissarsiassat inuuniarnermi imarmiunik tapertaqarnissamut inoqatigiinnut pingaaruteqarluinnarnerat, aammalu ilaqutariit nunaqatigiillu akornanni ataqatigiinnermut tapersersoqatigiinnermullu arfanniarnerup pingaaruteqarnera. Tamatuma saniatigut atuakkap takutippaa piniarnerup, aalisarnerup, arfanniarnerlu ilanngullugu, pissarsiassatut iluaqutissat nunaqqatigiit inuuniarnerannut qanoq pingaaruteqartigisut.

Index

UNIVERSITY PRESS OF NEW ENGLAND

publishes books under its own imprint and is the publisher for Brandeis
University Press, Dartmouth College, Middlebury College Press, University of
New Hampshire, Tufts University, and Wesleyan University Press.

Library of Congress Cataloging-in-Publication Data

Caulfield, Richard A.

Greenlanders, whales, and whaling : sustainability and self-determination in
the Arctic / Richard A. Caulfield.

p.   cm. — (Arctic visions)

Includes bibliographical references and index.

ISBN 0–87451–810–5 (cl : alk. paper)

1. Inuit—Fishing—Greenland.   2. Inuit—Greenland—Government
relations.   3. Inuit—Greenland—Economic conditions.   4. Whaling—
Government policy—Greenland.   5. Baleen whales—Government policy—
Greenland.   6. Sustainable fisheries—Greenland.   7. Human ecology—
Greenland.   8. Self-determination, National—Greenland.   9. Greenland—
Politics and government.   10. Greenland—Economic conditions.   I. Title.
II. Series.

E99.E7C38   1997

338.3'7295'09982—dc21   97–522